RFID Labeling

Smart Labeling Concepts & Applications for the Consumer Packaged Goods Supply Chain

D1308410

by Robert A. Kleist, Theodore A. Chapman, David A. Sakai, Brad S. Jarvis

PRINTRONIX®

Printronix, Inc.
14600 Myford Rd.
P.O. Box 19559
Irvine, CA 92623-9559

www.printronix.com

First edition, August 2004
Printed in the United States of America
Standard Book Number: ISBN 0-9760086-0-2
Library of Congress Catalog Control Number: 2004095650

Design: Eaton & Associates Design Company, Minneapolis

Contents

Foreword

Small in proportions, but monumental in its effects, RFID represents a big change in the retail industry. The recent initiatives of Wal-Mart, Target, the Department of Defense and other organizations, requiring case and pallet labeling for RFID, may seem trivial to the general public. That viewpoint is not shared by the relatively few people tasked with the job of bringing their companies into compliance over the next year or so. You may be affected already, perhaps as a member of your company's RFID task force, or as an employee of your company's supply chain.

RFID implementation is a real opportunity for those of us who support the retail industry. It is no small task, and we've observed a real lack of practical information. That is the reason for this handbook. Printronix is deeply involved in helping our clients meet RFID requirements. *RFID Labeling: Smart Labeling Concepts and Applications for the Consumer Packaged Goods Supply Chain* is part of our effort to help companies get started with the technology.

This book is a collaboration of information from our employee subject matter experts and business partners. The first five chapters cover the basics, with plenty of cross-references to later chapters that cover topics in more detail. We have made every attempt to provide accurate information, but we freely admit that neither we nor anyone else today are experts in all aspects of RFID. In fact, the speed of change in RFID technology for retail supply chains makes publishing a book on the subject a risky undertaking, and for this we ask your indulgence.

The subtitle of this book makes clear our point of view: smart labels offer the lowest cost, most practical, least disruptive way to implement RFID in the retail supply chain. With smart labels, you can succeed at achieving compliance with no business interruption. With smart labels, you can stream RFID into your current bar code labeling system, using proven tools to integrate both the physical tags and the associated EPC data, and even capture a database of information that provides traceability to each printed label.

Printronix has a rich 30-year history of enabling global printing solutions for supply chains. Companies worldwide use our thermal printers to produce bar code labels that guarantee 100% readable labels for retail packaged goods and government mandated identification requirements. We realize that printers are only an element of a package labeling process, so we collaborate with partners and industry groups to integrate seamless mission critical solutions. Our successful heritage includes many years of matching on-demand printing technologies to legacy applications and global enterprise management systems. We recognize that our solutions are only as good as the benefits that they bring to your supply chain, through increased efficiencies, the elimination of rework, and the availability of accurate and useful data.

By attaching smart labels on cases and pallets, retail packaged good companies are placing themselves at the forefront of a new era. What may appear to be a forced change with no return on investments may actually become a way for companies to rethink and reengineer processes, enhance their value as business partners, and capture a profitable return on your RFID investments.

As a member of EPCglobal, AIM Global Vendor Compliance Federation and the MIT RFID Packaging Special Interest Group, Printronix looks forward to helping you meet the goals of your RFID program. We would like to hear from you. Please take a moment to fill out and mail the reply card that you'll find inside the book.

August, 2004

Acknowledgements: the authors wish to thank the following people for their significant contributions to this book: Carol Ballesty, Scott Begbie, Bob Crum, Tim Eaton, Andy Edwards, Rick Fox, James Harkins, George Harwood, Jerry Houston, Karen Jensen, Van Le, Jim McWilson, Guy Mikel, Steve Morris, Christine Pruett, Terry Pruett, and Lisa Reickerd.

Definitions

Absorption
The degree to which materials change radio waves (electromagnetic energy) into current and heat.

Active tag
RFID tags having an active on-board transmitter, usually powered by battery, that constantly emits a signal, with a read range of 100 feet (30 m) or more. EPC Classes 3 and 4.

AIM-USA
Automatic Identification Manufacturers, a USA trade association.

Air protocol
The radio communications specification for a tag and reader, that describes the operating frequency (UHF 915 Hz), call and response characteristics (AM, half-duplex, pseudo-random frequency hopping, etc.), allowable transmission distances and applicable regulations.

Amplitude Modulation (AM)
Method of combining an information signal and an RF carrier, where a different voltage level is assigned to a digital 0 and 1.

Antenna
A radio frequency transducer. A receiving antenna converts an electromagnetic field into an alternating current. A transmitting antenna converts AC to an EM field.

Antenna detuning
Relative to RFID implementations, the reduction in the amount of energy that is available to power a tag, or the reduction of the size of the signal reaching the reader, due to the environment.

Anti-collision
See collision avoidance.

ASN
Advance Shipping Notification, used by Wal-Mart and others to confirm shipment in advance from a supplier.

Attenuation
The loss of energy as a signal propagates outward.

Auto-ID Center
Formed at MIT in 1999 to create standards and methods for RFID, the center closed operations and passed its work to EPCglobal.

Auto-ID tags
A general term for RFID tags used in the supply chain.

Backscatter reflection
Far-field electromagnetic waves that bounce off of and propagate away from an object. With passive tag RFID, the tags modulate the backscatter reflection to create a unique response.

Bar code
An automatic identification technology that encodes information into an array of adjacent varying width parallel rectangular bars and spaces.

Beam power tag
Another term for a passive tag, used in reference to the tag's power source, which is the RF energy "beam" coming from a reader.

Case
General term for product packages in trade item quantities ready for shipping.

Certified label
Labels that are approved by a printer supplier for use with their equipment.

Circular polarization
Antenna design where the energy is broadcast in a number of angles to its plane, creating a more circular pattern.

Collision avoidance algorithm
RFID reader firmware that intercepts multiple simultaneous tag signals, sorts responses, and initiates a communications protocol to sequentially collect the information.

Compliance
Term to describe planned activities that meet a mandate or directive.

Conductivity
The ability of a substance to carry an electrical current. Conductive materials such as liquids and metals tend to absorb radio waves (electromagnetic energy), and attenuate them.

CPG
Consumer Packaged Goods.

CRC
Cyclic Redundancy Check. The checksum calculation field in an EPC Class 1 tag.

Curtain
RFID reading area on a packaging line, where reader antennas are arranged and focused to pick up tags passing between them.

DC
Distribution Center.

Dead zone
General term for an area where an RF signal cannot be read.

Dielectric loss
Losses due to the non-conductivity of a substance.

DoD
Department of Defense.

EAN
European Article Number. An eight or 13-digit code originally used by companies outside North America to uniquely identify themselves and their products worldwide.

EAN International
Based in Brussels, EAN International is a member organization that jointly manages the EAN.UCC system with the UCC.

EAS
Electronic article surveillance. A theft detection system where RF tags are attached to high value clothing and other items in a retail store.

ECCnet
A service of the Electronic Commerce Council of Canada, providing a secure, online, single source of standardized item data continuously synchronized with trading partners.

EEPROM
Electrically Erasable Programmable Read-Only Memory. A non-volatile storage method used in some passive tags.

Electromagnetic field
Produced when charged objects such as electrons in a wire, accelerate and decelerate. All EM fields display properties of wavelength and frequency.

EMI
Electromagnetic interference. Energy byproducts of motors, sunspots, etc., within the RF spectrum that disrupt communications.

Encoder
A reader and antenna built into a smart label printer to write information to tags.

EPC
Electronic Product Code. A labeling code that identifies the manufacturer, product category and individual item. Created by the Auto-ID Center/EPCglobal, EPC is backed by the United Code Council and EAN International, the two main bodies that oversee bar code standards.

EPCglobal
EPCglobal is a joint venture of the EAN Internatio nal and Uniform Product Council, representing 100 member organizations worldwide.

EPC Header
Identifies the EPC version number and allows for the evolving of different lengths or types of EPC (8 bits).

EPC Manager
Identifies the manufacturer of the product (28-bits).

EPC Network
A global information repository, based on the Internet, and administered by VeriSign, that serves as a registry of EPC numbers and associated information.

EPC Object Class
Refers to the exact type of product (24 bits).

EPC Serial Number
The unique identification number for the item (36 bits).

ERP
Enterprise Resource Planning. Usually refers to a software application.

Far field
The distance at about one wavelength where the electromagnetic field separates and radio waves propagate away from the magnetic (near) field.

Field strength
A measure of radio signal reception.

Firmware
Software programmed into a non-volatile memory chip.

Fixed RFID reader
A reader installed at a location, with an external power supply, antenna and network connection.

Frequency
The number of repetitions of a complete waveform in a specific period of time. 1 kHz equals 1,000 complete waveforms in one second. 1 MHz equals 1 million waveforms per second.

GIAI
Global Individual Asset Identifier. Used by a company to label fixed inventory.

GLN
Global Location Number. Used within the EAN/UCC-13 data structure to identify physical, functional and legal entities.

Global data synchronization
A multi-industry objective and process whereby item data is represented in standard formats, and these formats are universally adopted to facilitate electronic information exchange between manufacturers, distributors and retails.

Good label
A label is considered good when the RFID data is written to the tag correctly, the correct image is printed and content data is verified against the source.

GPIO
General Purpose Input Output interface for Printronix devices.

GRAI
Global Returnable Asset Identifier. Used typically to track returnable containers.

GTAG
Global Tag. An EAN.UCC initiative supported by Philips Semiconductor and others for asset tracking and logistics.

GTIN
Global Trade Item Number. An umbrella term used to describe the entire family of EAN/UCC data structures for trade items.

Half-duplex
Mode of radio communication where the same band is used both ways and transmission occurs in one direction at a time.

Handheld RFID reader
Battery powered reader used to verify tags and locate tagged items.

Header
The beginning portion of an EPC number, used to indicate the version.

Hertz
A measure of electrical frequency oscillation.

High frequency tag
Tags operating in the 13.56 MHz band.

Host computer
Computer that runs software that interacts with RFID and other devices, such as a warehouse management system.

IC
Integrated Circuit.

Inductive coupling
Method of creating a current in a conductor without touching it directly to a power source. A tag responds to a reader by inductively coupling with the reader carrier signal.

Inlay
The combined chip and antenna mounted on a substrate and attached to label stock to create a smart label.

Interference
Any environmental condition that creates electrical noise at the same frequency as the communications signal.

Interrogator
Another term for a reader.

ISM bands
Industrial, Scientific and Medical government-regulated radio frequency bands.

Item
General term for an individual product or service.

Linear polarization
Antenna design where energy radiates mostly in a straight line pattern at 90 degrees to its plane.

Line-of-sight
Technology that requires an item to be "seen" to be automatically identified by a machine. Bar codes and optical character recognition are two line-of-sight technologies.

Low-frequency tag
Tags that operate in the 125 KHz band.

Microchip
A microelectronic semiconductor device comprising many interconnected transistors and other components. Also called an IC.

Microwave tag
Tags that operate in the 5.8 GHz band.

Nano-technology
Integrated circuit technology where circuit size can be expressed in nanometers, a unit of which is only 3-5 atoms wide.

Near field
Portion of the electromagnetic field, within a distance of one wavelength, where the magnetic field follows a flux line path.

Network print management
A system of printer/encoders communicating through a host computer interface, allowing centralized control and visibility into the enterprise printing function.

ONS
Object Name Service. Similar to the Domain Name System associated with the World Wide Web, ONS is an Auto-ID Center designed system for looking up unique EPC through an Internet linked computer.

ODV™
Printronix Online Data Validation.

Opaque
Relative to RFID, an opaque object does not allow a reader signal to pass through it.

Order cycle time
The speed at which a retail location can replenish stock. Used as a measure of supply chain efficiency.

Out-of-stock
Retail term for product offered for sale but not in inventory.

Overstrike
Printer capability to recognize a bad label, back up and mark it as such by printing a grid on top of it.

Pallet
General term for a skid full of cases that can be handled by a fork lift.

Parametric test
Test of a tag inlay after assembly to detect quiet tags.

Passive RFID tag
Passive tags do not have an on-board powered transmitter. They are activated by the electromagnetic waves of a reader, with a read range of 10-25 feet.

PGL™
Printronix Graphics Language. Native language of Printronix printers, for programming graphics and control commands.

Pilot
The stage of an implementation where technology, process and methods are evaluated before committing to their use in a production environment.

PML
Physical Markup Language. An Auto-ID Center designed method of describing products in a way computers can understand. Based on XML.

Polarization
The orientation of flux lines in an EM field.

Portal
A defined RFID reading area, where readers are mounted specifically to read tags going through, such as a dock door or over a packaging line.

POS
Point Of Sale.

Print and apply
Automated approach to printing, encoding and applying smart labels in a packaging process.

PrintNet® Enterprise
Printronix networked print management system.

Print quality
The measure of compliance of a bar code symbol to AIM requirements of dimensional tolerance, edge roughness, spots, voids, reflectance, quiet zone and encodation.

Quiet label
A label that cannot be read from a normal distance.

Radiolucent
The degree to which a substance absorbs or transmits radio waves that attempt to pass through it.

Radio waves
Waves at the lower end of the electromagnetic spectrum.

Read after print
A step in producing smart labels, where the printer interrogates the RFID tag and reads the bar code to verify a good label.

Reader
Also called an interrogator. The reader communicates with the RFID tag and passes the information in digital form to the computer system.

Read only memory (ROM)
A form of storing information on a chip that cannot be overwritten.

Read only tag
Encoded during the tag manufacturing process so information cannot be changed. EPC Class 0.

Read range
The distance from which a reader can communicate with a tag. Range is influenced by the power of the reader, frequency used for communication and the design of the antenna.

Read rate
The accuracy of a bar code or RFID system, expressed as a percentage of good reads versus bad. May also be a benchmark or theoretical rate.

Read redundancy
The number of times a tag can be read while in the read window.

Read write tag
A programmable tag, EEPROM or battery-backed memory. EPC Classes 1-4, including UHF Gen 2.

Receiver
Another term for a reader, as in radio receiver.

RIED
Real-time In-memory Event Database.

Replenishment
Supply chain term for ordering and receiving stock.

RF
Radio Frequency.

RFID
Radio Frequency Identification. A method of tracking using radio waves that trigger a response from a device attached to an item.

RFID Printer/encoder
An RFID printer that encodes the smart label and immediately checks to verify if the tag is readable. Same as smart label printer.

ROI
Return on Investment, usually measured as capital equipment and labor costs offset by tangible and intangible returns over a given period of time.

Savant
Auto-ID Center term for distributed network software that manages and moves data related to Electronic Product Codes.

Semi-passive tag
Passive tag that uses battery-power to boost its transmission range.

Serialization
The unique numbering of objects for identification purposes.

Shielding

Materials that have a significant electrical conductivity. Shielding may be used to directionally orient a signal, such as the wire braid jacket on an antenna cable. Objects that are electrically conductive at high frequencies, such as aluminum cans, may cause shielding that is detrimental to the performance of RFID.

Shrinkage

Retail term for product loss after receipt from a supplier, usually due to employee theft.

SKU

Stock Keeping Unit.

Slap and ship

Manual approach to applying RFID tags or smart labels to packages in a distribution center.

Sleep mode

An ID tag command to suppress an RF response in a tag. Used to minimize unintended reads.

Smart label

A label that contains an RFID tag. It is considered "smart" because it can store information, such as a unique serial number, and communicate with a reader.

Smart label printer

RFID printer that produces smart labels.

Squiggle tag

An RFID tag made by Alien Technologies with a squiggle-shaped antenna.

SSCC

Serial Shipping Container Code. Used to identify cases, pallets and other containers.

Strap

Section of an RFID chip surrounding the chip, to which the antenna is attached.

Supply chain

Industry term for a group of companies working together to manufacture, inventory and supply materials to fulfill a production schedule or finished products to meet market demand.

Tag

The generic term for a radio frequency identification device.

Thermal printer

A device that uses a heat transfer process to print characters or graphics, for continuous roll printing of tape or labels.

TMS

Task Management System.

Tote

A re-usable container that can be hand-carried.

Trade item

General term for a product or products packaged in lot sizes and make available for transport and sale.

Transmitter

Portion of a reader that includes the antenna and is used to broadcast a radio signal.

Transponder

Refers to the part of an active (battery-powered) tag that includes the antenna and is used to emit a response to a signal.

UCC

Uniform Code Council. Based in the United States, UCC is a membership organization that jointly manages the EAN.UCC system, including the Uniform Product Code in the USA and Canada.

UCCnet

An Internet-based product registry service for standards-based electronic commerce. The service enables synchronization of item and location information among trading partners.

UHF

Ultra-High Frequency. The term generally given to waves in the 300 MHz to 3 GHz range. UHF offers high bandwidth and good range, but UHF waves don't penetrate materials well and require more power to be transmitted over a given range than lower frequency waves.

UHF Generation 2 Foundation Protocol

Formal name for the emerging EPC class standard for passive RFID tags. UHF Gen 2 is expected to replace the current class 0 and 1 specifications.

UID

Universal Identification, code used by USA Department of Defense to mark and track assets.

UPC

Universal product code. The bar code standard used in North America over the last fifteen years. It is administered by the Uniform Code Council.

Wavelength

The inverse of frequency, a wavelength is the measure of peak to peak of a radio wave.

WMS

Warehouse Management System.

WORM tag

Write Once, Read Many tag, using a type of non-volatile memory that can be written to only once, typically just before it is applied to product or container. Thereafter the information is fixed and can be read only. EPC Class 0+, 1 and UHF Gen 2.

XML

Extensible Markup Language, for defining, validating and sharing documents containing structured information. Unlike HTML, XML tags can be designed for specific purposes.

CHAPTER 1

Introduction

When Wal-Mart announced in late 2003 an initiative for RFID use within its supply chain, suddenly the rules of the game changed. Wal-Mart's top 100 suppliers were being asked to invest and re-engineer in order to retain their status as suppliers. Other US and European retail companies started similar RFID initiatives, including Albertsons, Carrefour, Metro, Target, and TESCO. The US Defense Department issued a similar directive. Before anyone could get used to the idea, thousands of companies fell under some sort of RFID compliance initiative.

It is fair to ask what got us here. Why did all of these companies get served with a mandate? What's wrong with their current bar coding system? What is it about RFID that is so compelling that Wal-Mart, the DoD and others will risk potential disruption within their supply chain? After all, this isn't just slapping a tag on a case or pallet. In order to implement RFID, you must have systems in place to serialize tags on cases, pallets and eventually every item you produce and ship. Then you need systems to read, track and leverage value out of all of that data. Finally, you need systems to synchronize data throughout the supply chain. That's a huge change.

Can RFID help companies solve real business problems? Can it truly improve inventory tracking, product shrinkage and out of stock at the retail store level? The basic premise of RFID is that by attaching a radio frequency tag to an object, a computer can track that object without human intervention. By tracking an object remotely through key events in that object's "life," you can automate its flow through the supply chain. Business rules written into ERP and WMS software can guide its flow, from raw materials to retail shelf, from factory to fox hole. Presumably, RFID will help supply chains tune themselves to respond more quickly to consumer demand.

If RFID can do this, then compliance is suddenly not the issue. Business survival is the issue – staying ahead of the curve, the coming revolution, where only the RFID-enabled supply chain companies succeed.

REVOLUTIONARY TECHNOLOGY, OR COST BURDEN?

The question is not whether RFID is disruptive, rule-changing technology. Compliance mandates took care of that. Question is, can RFID deliver enough economic benefits to transform the way supply chains do business and recoup the upfront costs? Will this be true for every supplier?

Like the personal computer, the fax machine, the Internet, and bar codes in their inception, RFID has the potential to transform commerce. RFID promises to take people out of the identification and data collection loop. A recent study by IBM estimates that RFID could reduce labor involved in the receipt of goods by 60 to 90 percent. By helping companies better track, automate the flow of and understand the condition of goods in the supply chain, it improves product availability at the retail level without adding inventory. Some experts estimate that 30 percent of inventory in the supply chain is buffer stock – it exists because demand and supply information is not precise and real-time. A recent consumer packaged goods study estimates that $5 billion of obsolete goods are written off each year. Wal-Mart estimates that RFID will allow it to recapture 1 percent of revenue by improving out-of-stock, which translates to about $2.5 billion. A.T. Kearney estimated the average out-of-stock improvement using RFID to be 3-5%.

Its clear that labor savings is not where the bigger benefits of RFID lie; rather, bigger benefits have to do with its potential for solving a data-availability problem. Trouble is, RFID will create huge amounts of data that is not necessarily accurate, appropriate, actionable, complete, compatible, or unique. All the data issues have to be solved to gain the benefits. Most companies still cannot accurately forecast future requirements, despite having ample data resources available to them through their supply chain partners.

Several convergent factors determine the timing of technology adoption, and help answer the question "why RFID now?":

Standards – Global, pan-industry standards-making organizations are taking an active role. There is strong consensus, although there are many issues yet to be ironed out, and it's probable that standards variations will exist for the foreseeable future.

→ See Page 43 for RFID standards.

Costs – Costs for tags and readers are declining and are expected to decline even further in the coming years.

Multiple suppliers and vendors – A number of well-funded product companies, with supply chain expertise, have announced products. There is competition in the market for both products and services for system integration and achieving compliance.

Leadership – Wal-Mart, DoD and other leading companies and opinion leaders are committed to RFID. These leaders are asking their supply chains to formally commit to RFID through their compliance initiatives.

→ See Page 89 for Wal-Mart & DoD initiatives.

BUILDING A BUSINESS CASE FOR RFID

For a company that supplies to a major retailer or the DoD, the case for implementing RFID may appear straightforward, but that's only on the surface. The business case has many layers. It is a case for the boardroom as well as the shipping dock. The value proposition is different for every company. Not all companies will benefit from RFID equally. Some will see significant returns. Others will see very few, if at all in the near term. Issues for suppliers to consider include:

Risk of non-compliance – Figure 1.1 illustrates some of the issues. Near term, you stand a chance of losing business if your customer insists that you comply with an RFID mandate and you don't. If the customer represents a large percentage of your business, that is sufficient justification for some level of compliance activity. If the customer is a relatively small part of your business, the risk of non-compliance may be acceptable when compared to the cost of implementing RFID. The yearly profit of that piece of business may not be anywhere near the up front costs of RFID. You could elect to forego that business in the near term, implement RFID if necessary at some point in the future, and risk later getting back in good graces with your customer. The issue changes, however, if other companies in the retail industry and elsewhere announce RFID mandates, and you are a supplier to them as well.

High-volume, low-margin product businesses – Examples of such products are soaps and cleaners, frozen foods, perishables, everyday wear, etc. If a high percentage of your business is affected, and you produce or re-package low-margin products in high volume, the cost of compliance is considerable, and could effect your overall profitability of doing business. An RFID tag will be a significant cost adder to each case of product shipped (Fig. 1.2). A business case could be made to implement ways to achieve direct labor savings and internal efficiencies from RFID as soon as practical. On the other hand, if a relatively small portion of your business is affected, a business case could be made to reducing exposure to RFID implementation costs by outsourcing the RFID compliance

(→) SEE FIG. 1.2

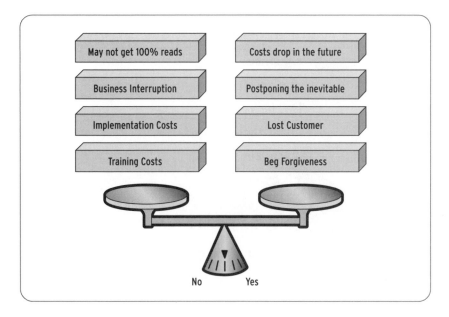

May not get 100% reads

Costs drop in the future

Business Interruption

Postponing the inevitable

Implementation Costs

Lost Customer

Training Costs

Beg Forgiveness

No   Yes

FIGURE 1.1

Assessing the risk of non-compliance.

work, or diverting production to an area of your distribution center where it can be re-worked by hand. By avoiding the high costs of re-engineering all your processes for RFID in the near term, you may gain in the long run should implementation costs drop drastically as RFID moves toward mass adoption.

Low to medium volume high-margin product businesses – Examples include bicycles, cosmetics, video games, electronics, etc. – For computer peripherals, televisions, and other big-ticket products, where the case equals the item-level, RFID and EPC implementation could bring near immediate benefits. Item level serialization is already being done, and RFID tracking could

FIGURE 1.2

Tag cost as a component of product cost.

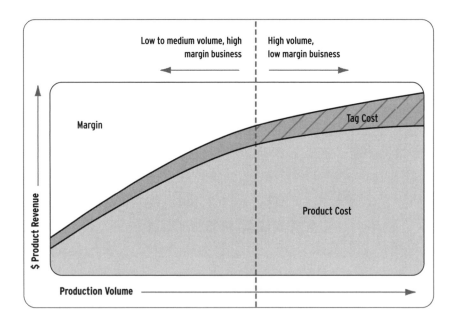

improve out-of-stocks and item shrinkage, which are significant cost issues. See Figure 1.3. Internal implementation, rather than outsourcing to a contract RFID compliance center, may be the best choice. A number of trial options are available, including manual tagging and diverting product to semi-automated lines. By staging the implementation, a company can advance on the RFID learning curve as it works toward compliance.

Complex supply chain – Companies that supply products direct to store, such as perishables (flowers, bakery goods), contract bottlers or jobbers for snack items, should look to their customer and supply chain partners for assistance. It may be possible to install RFID tags

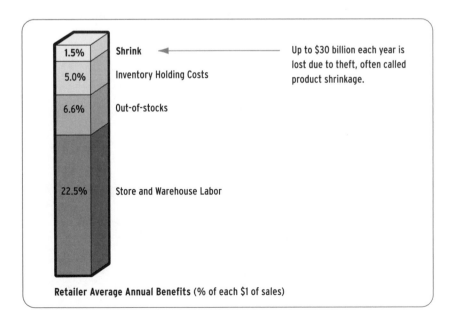

1.5% Shrink — Up to $30 billion each year is lost due to theft, often called product shrinkage.

5.0% Inventory Holding Costs

6.6% Out-of-stocks

22.5% Store and Warehouse Labor

Retailer Average Annual Benefits (% of each $1 of sales)

FIGURE 1.3

Big-ticket items are more affected by out-of-stocks and shrinkage.

on re-usable containers such as totes and pallets. Direct to store supply chains may require significant cost sharing among partners.

Pushing RFID down the supply chain – A business case for RFID may require a hard look at your own supplier practices. Depending on your business, the direct benefits "within the four walls" may not be as significant as those gained by having your suppliers and trading partners implement RFID (Fig. 1.4). If you are a volume-driven manufacturer who dominates your industry, you stand to gain considerably by pushing RFID down to suppliers.

It is fair to say that benefits of RFID will not accrue at the same rate, and in the same fashion, in each link of the supply chain. The organization at the top of a supply chain – the retail store, the

FIGURE 1.4

RFID benefits gained through supply chain and trading partners.

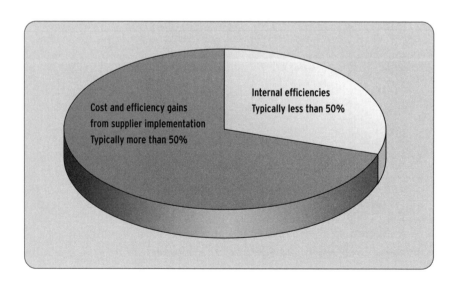

Internal efficiencies
Typically less than 50%

Cost and efficiency gains
from supplier implementation
Typically more than 50%

troops at the front in the case of a military supply chain – will gain most quickly by elimination of out-of-stocks. Other suppliers in the chain, however, will also see gains through increased revenues by having their products available on the retailer's shelves. Companies that produce big-ticket items, manage high value assets, or control the point of sale, and that implement RFID systemically, will benefit quickly as well. Everyone else will have to wait for the trickle down of transparent, instantaneous data in the supply chain, and the labor savings as RFID pilots mature. It might make sense to consider RFID in two time frames:

Near term, compliance driven – Integrate RFID into your case and pallet supply systems with as little downtime and business interruption as possible. One obvious path is by converting your existing UPC bar code labeling stream to EPC smart labels.

→ See Page 89 for compliance requirements for Wal-Mart and DoD initiatives.

Long term, ROI driven – Use RFID to change the rules in your own supply chain and business. Maximum value will come when you've taken the opportunity to re-think your processes. Look for artifacts of legacy material handling systems based on old technology and restructure them. Eliminate procurement and manufacturing planning based on inventory buffering and cueing. Comprehensive changes such as these should only be undertaken when you have internalized all the challenges and potential obstacles for successful RFID technology implementation, and can engineer appropriate solutions.

PROTECTING INVESTMENTS DURING IMPLEMENTATION

Any business case for early adoption will be helped by the RFID team making smart choices about approaches, equipment and vendor partners. Equipment and system integration will be the two largest investments. If they are kept to one-time costs, returns tend to accrue sooner. Here are some guidelines to consider:

Upgradeable devices – RFID printers and readers that are firmware upgradeable will retain their use through the lab, pilot and production phases of an implementation. Look for devices that are engineered specifically for rugged use in production environments, device suppliers with proven industry experience and a commitment to helping you protect your investment. This is especially important as compliance requirements shift to the UHF Gen 2 standard in 2005.

Software migration tools – Device vendors should provide migration tools that support the conversion of data streams set up for UPC bar coding over to EPC. This includes capabilities to manage the assignment of EPC serial numbers across multiple operations producing smart labels.

Certification of labels and tags – As the single biggest cost item, you will need flexibility in sourcing labels. Look for vendors who offer label compliance and certification services, and professional services teams who will work with you directly to help you achieve compliance.

System integration and automation – Vendor teams who have RFID and supply chain experience working together on similar projects offer the best assurance of success as you move toward volume production. Many of these system integration teams tend to be from smaller firms that specialize in retail packaging applications. Their experience and ability to work with both legacy systems and open-source languages and protocols will help ensure against having to scrap a system and start over.

Backup and recovery capability – Smart labels with bar coding offer the best way to identify and disaster recover from RFID problems. Look for ways to stream RFID into your current bar coding process, and partners that will help you retain the integrity of both systems for the foreseeable future.

HOW THIS BOOK IS ORGANIZED

This book covers a number of related subjects important to understanding and using RFID and smart labels in particular. It is not necessary to read each chapter in sequence. Chapters are organized in subjects that cover various layers of an RFID system, from the device layer (Chapter 2), to the systems integration layer (Chapter 9). Figure 1.5 shows the layers and related chapters. At the back of this book is a supplementary section of products from Printronix.

→ SEE FIG. 1.5

Sources and Further Reference

Meeting the Retail RFID Mandate, white paper, A.T. Kearney, November 2003.

Corporate Strategy for the New Millennium, white paper, IBM, available: www.ibm.com.

FIGURE 1.5

Organization of the book.

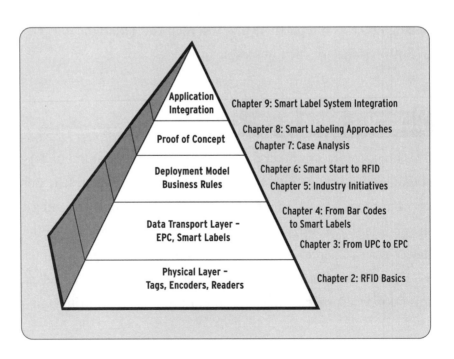

CHAPTER 2

RFID Basics

A typical RFID system consists of four main components: tags, an encoder, readers, and a host computer. See Figure 2.1. The RFID tag is made up of a microchip and a flexible antenna encased in a plastic-coated inlay. The encoder is used to write information to the tag. In coming years you might find RFID tags built into products and product packaging. For now, the most common format is a shipping label with a built-in tag, or smart label. Smart labels can be printed and placed on each case or pallet.

To get an RFID tag to "talk back" a reader broadcasts a radio wave. If it is within the range of the reader, the tag answers, identifying itself. Tags can be read from a distance without physical contact or line of sight. The distance within which a reader can communicate with a tag is called the read range. Communications between readers and tags are governed by protocols and emerging standards, such as the EPC UHF Class 1 standard for supply chain applications.

RADIO WAVES

When electrons move, they create electromagnetic waves that can move through air. The waves can pass through some physical objects, and even a vacuum. The number of oscillations per second of an

FIGURE 2.1 Main components of an RFID system.

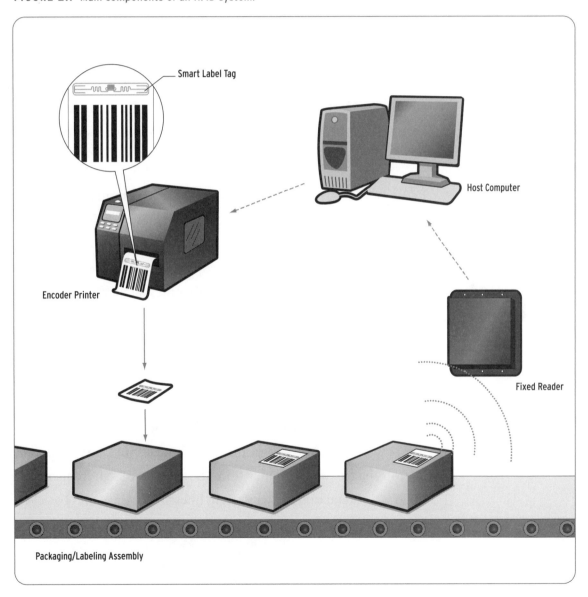

Smart Label Tag

Host Computer

Encoder Printer

Fixed Reader

Packaging/Labeling Assembly

electromagnetic wave is called its frequency, and is measured in Hertz (Hz). The distance between two consecutive wave peaks is called the wavelength.

By attaching an antenna of the appropriate size to an electrical circuit, the electromagnetic waves can be broadcast efficiently and received by a receiver some distance away. All wireless communication is based on this principle.

Radio is usually associated with long distance communications. In the case of RFID, we focus on the characteristics of radio waves over a relatively short distance. Electromagnetic waves travel through a vacuum at the speed of light. In copper wire, the speed slows to about two-thirds of this value and becomes somewhat frequency dependent. The electromagnetic spectrum is shown in Fig. 2.2. The radio, microwave, infrared and visible light portions of the spectrum can all be used for transmitting information by modulating the amplitude, frequency or phase of the waves.

The properties of radio waves are frequency dependent. At low frequencies, radio waves pass through obstacles well, but the power falls off sharply with distance from the source. At high frequencies, radio waves tend to travel in straight lines and bounce off obstacles. They diffract at corners, sharp edges, and openings.

Radio waves are subject to interference from a variety of sources, from sun spots to other electrical equipment. Radio communications has its uses in all sorts of applications, but only if interference between users can be kept at a minimum. For this reason, governments tightly license users of radio transmitters, with the exception of the industrial, scientific and medical (ISM) bands. Transmitters using these bands do not require government licensing. One band is allocated worldwide: 2.400–2.484 GHz. In addition, in the United States and Canada, bands also exist at 866–956 MHz and 5.725–5.850 GHz. These bands are used for cordless telephones, garage door openers, wireless hi-fi speakers, security gates, etc. See Table 2.1.

 **SEE TABLE 2.1**

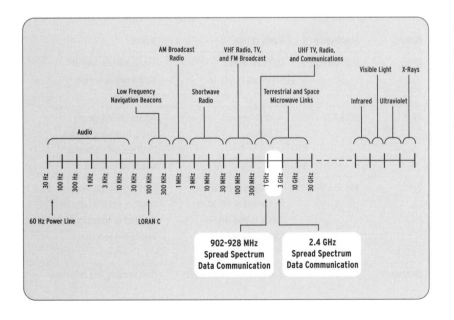

FIGURE 2.2
Portion of the electromagnetic spectrum associated with RF communications.

Many parts of the world have specific frequencies assigned to various applications, from cell phones to security gates. This hinders the development of a single global standard for supply chain RFID use. In the United States, the FCC has long authorized the 915 MHz band for RFID use. Recently, Europe has adopted 866 MHz for their RFID standard. Japan is currently considering standardization at 950-956 MHz.

THE RFID TAG

→ SEE FIG. 2.3

An RFID tag is made up of a microchip anchored to a strap, which is attached to an antenna, and encased in a protective inlay. See Figure 2.3. The actual chip may be no bigger than a grain of sand

TABLE 2.1

For RFID, these bands have the following characteristics:

Band	Frequency	Read range	Notes
LF	100-500 kHz	Up to 20 inches (50.8 cm)	Access control, animal identification, vehicle key-locks.
HF	13.56 kHz	Up to 3 feet (1 meter)	Access control, smart cards, item level tagging, libraries, and electronic article surveillance.
UHF	866-956 MHz	FCC allows over 20 feet (6 meters) at 915 MHz. Range at 866 MHz is about 10% less than at 915 MHz.	Supply chain use, baggage handling and toll collection. Wal-Mart is accepting RFID tags in this spectrum.
Microwave	2.45 GHz	3 to 10 feet (1 to 3 meters)	Item tracking, toll collection.

(about 0.3 mm^2). Chips used in RFID tags may become the most widely-used commercial application of nano-technology. Although the chips are tiny, the antennas are not. They need to be big enough to pick up a signal. The antenna allows a tag to be read at a distance of 10 feet (3 meters) or more, and through many materials including boxes. Antenna size tends to determine the size of an RFID tag. Once an antenna is attached and the assembly is packaged in a protective laminate, the resulting RFID tag becomes finger size or larger, at least for now.

Figure 2.4 illustrates a typical tag IC circuit design. The low-power circuits handle power conversion, control logic, data storage, data retrieval, and modulated backscatter to send data back to a reader.

→ **SEE FIG. 2.4**

Figure 2.5 shows examples of tags having different antenna designs optimized for various applications. Antennas can be made of silver, aluminum and copper, and created by material deposition techniques similar to squirting ink onto a page. The amount of metal used and antenna size determines a tag's sensitivity. Sensitivity is crucial, since it has been found that everyday items, such as cases of bottled water or aluminum cans can attenuate RF signals.

→ **SEE FIG. 2.5**

Tag design, tag placement, case orientation, and reader location all play a role in achieving consistent performance. Tag antennas can be made in a variety of configurations to achieve various performance characteristics. They need to be an optimal fraction of

→ **SEE PAGE 135**
for characteristics
affecting read rate.

FIGURE 2.3

Tag IC and Antenna.

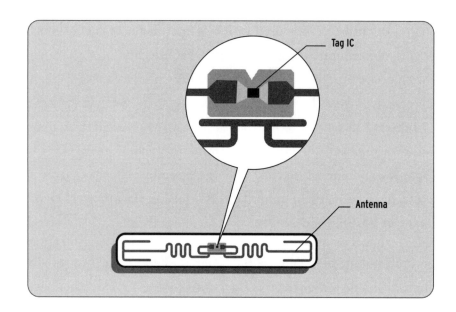

FIGURE 2.4

Typical tag IC circuit design. The low-power circuits handle power conversion, control logic, data storage, data retrieval, and modulated backscatter to send data back to a reader.

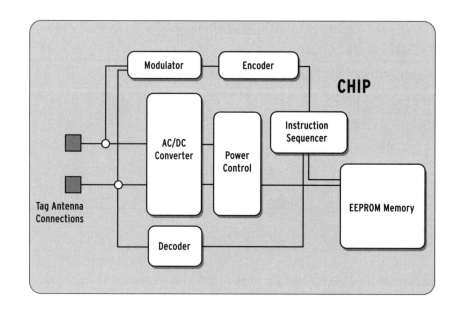

the operational frequency wavelength. A $1/4$ wavelength is typical. The 900 Hz wavelength is approximately 1.09 feet (33 cm). Some tag antennas are designed to address a broad range of conditions. Others are designed specifically for a narrow range of conditions such as for readability when affixed to metal packaging. Tag antennas may also optimized for reading by a specific type of reader, or with a reader antenna in a specific position.

RFID tag suppliers include:

- Alien Technology, www.alientechnology.com
- Avery Dennison, www.averydennison.com
- Impinj, www.impinj.com
- Matrics Systems Corporation, www.matrics.com
- Philips, www.semiconductors.philips.com
- Rafsec, www.rafsec.com
- Texas Instruments, www.ti-rfid.com

Tag development is in its early stages. The number of designs and manufacturers is certain to grow. As standards emerge, and adoption increases, you'll likely find alternate suppliers for the most popular tag types, and lower costs as volumes increase. The expected demand for tags is huge, considering that Wal-Mart estimates its annual case/pallet volume at greater than 8 billion units.

(!)
The expected demand for tags is huge, considering that Wal-Mart estimates its annual case/pallet volume at greater than 8 billion units.

FIGURE 2.5 Examples of various tag designs.

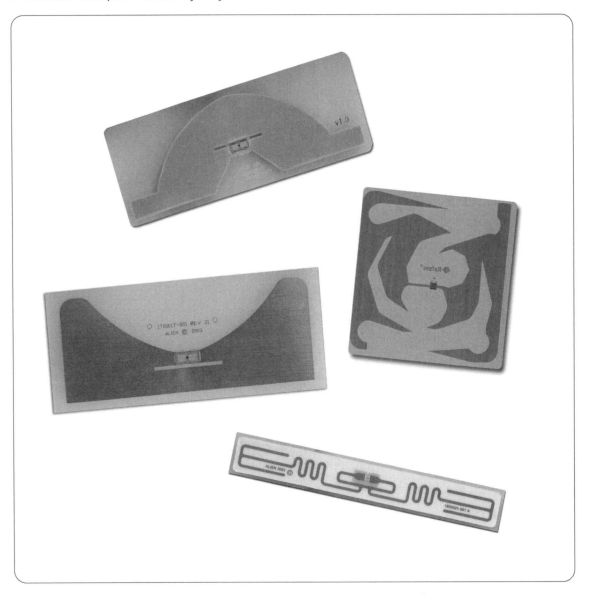

Tag Types

Tags for supply chain use come in a few basic types. One distinguishing characteristic is whether a tag is active or passive. Active RFID tags broadcast under their own power. An on-board battery runs the microchip's circuitry and transmitter. Active tags are capable of receiving and transmitting signals the distance of a football field. They are well-suited to applications where they can be permanently mounted and maintained, such on railroad cars for tracking movement in a switching yard, on ocean shipping containers, and on high-value military items stored in outdoor supply depots or bases.

Passive tags, on the other hand, have no battery. Instead, they draw power from the reader. Electromagnetic waves transmitted from the reader induces a current in the tag's antenna. The tag uses that energy to talk back to the reader. The "talk back" is known as backscatter reflection. See Figure 2.6. It is similar to how radar works. Whereas radar backscatter is more like an echo, the tiny circuit in an RFID tag can power itself with the induced current, and its backscatter is an amplitude modulated (AM) response. The AM signal can be interpreted as a digital signal of ones and zero's.

⊙ SEE FIG. 2.6

When they are not in the presence of a reader signal directed at them, passive tags are just that: passive – not capable of emitting any radio signal by themselves. They do not add unnecessary electromagnetic noise to their surroundings.

The majority of interest in supply chain RFID is centered on passive tag applications. Tag cost is one key reason. Tag cost is a consideration for supply chain applications even at the case/pallet level. Wal-Mart's estimated case pallet volume is 8 billion alone. A five-cent ($0.05 US) tag cost is widely considered to be the point at which mass adoption will be justifiable (Fig. 2.7). Tag costs are currently several times that number, but are expected to drop rapidly. For many applications, tags have considerable value even at their current cost, and the value will increase as tag usage grows.

FIGURE 2.6

Tag-Reader-Host communications.

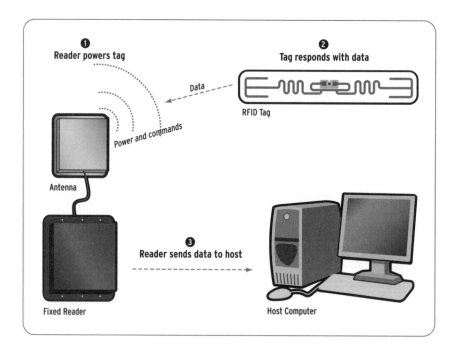

Passive tags are smaller in size, lighter in weight, have longer lives and are subject to less regulation than active tags. Passive tags operate only over relatively short ranges, and have limited memory when compared to active tags. Passive tags have difficulty performing in environments where a large amount of interference exists.

Another type of tag is the so-called semi-passive tag. It has many of the characteristics of a passive tag (small, lightweight, limited memory), but with battery backup to extend the answer range. Semi-passive tags are finding uses on shop floor containers and pallets for parts kitting and just-in-time manufacturing applications.

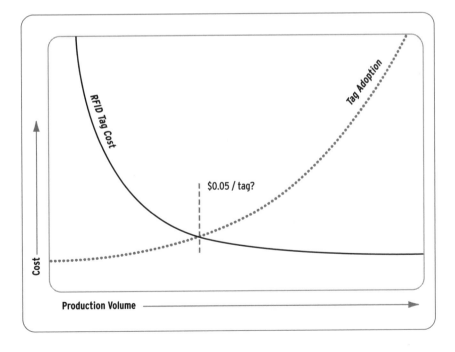

FIGURE 2.7

Tag costs will drop as production volume increases.

Tag Programmability

→ SEE PAGE 74
on encoding labels.

Another major way to distinguish tags is by their programmability. Chips in RF tags can be erasable and programmable (EEPROM), write-few-read-many (WORM) or read-only. Each re-write completely replaces the data in memory. The read range of a tag in most cases is farther than its write, or programming range. Most applications, therefore, will involve a tag encoder placed in-line with a tag to initially write data into it.

A read-only tag is preprogrammed by the circuit manufacturer. The information on these tags can never be changed. Pre-programmed tag information is linked to a product through a registry entry made in a warehouse management system (WMS) or host computer.

Table 2.2 summarizes the basic characteristics of various tag types. Right now it is hard to predict which types of tags will ultimately be adopted as a standard and what they will cost. Market requirements, order volumes, and the subsequent efficiencies gained by high volume manufacturing, rather than circuit features, may be the biggest cost determinants in the long run.

Tag Standards

EPCglobal, the RFID standards organization, is in the process of developing standards for RFID tags. These standards cover both the "air protocol," how tags communicate, and the programming technique used to store and read data. There are several classes of RFID tags in various stages of proposal and standardization. They range from simple read-only to more powerful tags that can broadcast their own signal and require their own power source.

The EPC version 1.0.1 tag standard is the current working specification for tags operating in the frequency range of 866–956Mhz. The intended open-air reading range of tags implemented to this specification is 6-10 feet (2-3 m), and no more than 34 feet (10 m) at their best orientation. Reader-to-tag and tag-to-reader communication occurs as half-duplex, like a walkie-talkie. The reader transmits a

→ SEE PAGE 80
on reading smart labels.

Tag Type	Advantages	Disadvantages	Application
Active	Greater read range, memory capacity, continuous signal.	Batteries require maintenance. Larger size.	Used with high-value asset tracking
Semi-passive	Greater read range, longer battery life	Battery wear and expense.	Reusable containers and asset tracking
Passive Read/Write	Longer life, multiple form factors, erasable and programmable	Time and expense to program.	Case & pallet applications. Approved for use with Wal-Mart
Passive WORM	Suited for item identification, controllable at the packaging source	Limited to a few re-writes, replacing existing data with new data.	Case & pallet applications. Approved for use with Wal-Mart
Passive Read Only	Simplest approach	Identification only, no tracking updates	Case & pallet applications. Approved for use with Wal-Mart

TABLE 2.2

Comparison of active and passive tags with read-only, read/write & WORM.

continuous wave signal. The tag modulates its backscatter reflection of the signal. The tags can be EEPROM, WORM or read-only.

Tag Selection and Readability

Product packaging and supply chain processes present a myriad of challenges for RFID. Products and product packages come in all sizes and types, and tags must be able to physically survive the crushing weight of a load. They must withstand shipping wear, temperature extremes, and material handling machinery.

Tag readability is also dependent on characteristics of the UHF spectrum. Here is a list, in no special order, of considerations when selecting a tag for supply chain use:

⊙ SEE PAGE 127
on case analysis.

- **Placement** – Read rate is affected by the orientation of the tag on a box or pallet relative to the reader.

- **Size and form factor** – Cases often have a specific place for a label, and some companies specify size and format.

- **Read speed** – The amount of time the tag is within read range in the case of a tagged carton on a moving conveyor, or a tagged pallet on a truck moving through a dock door portal.

- **Read redundancy** – The number of times a given tag can be read while in the reading area. If a tag can respond at least three times to read requests while it is in the reading area, chances are very good that its data will be captured without error.

- **Data requirements** – Tags will contain different information depending on their use (item, case, pallet, returns).

- **RF interference** – Read rates will be affected by sources of RF noise, proximity to other tags, and the composition of packaging materials and surrounding surfaces.

- **Harsh environments** – Steam, corrosive chemicals or extreme cold will affect the adhesive on a tag if nothing else.

- **Re-use** – Re-use could include use on re-usable containers, or as a way to document returned goods.

- **Cross-border shipping regulations** – Tags may have different read range and sensitivity depending on their frequency range of operation, due to different global standards.

- **Read speed** – The speed at which a tag will be expected to move through a reader array or portal.

- **Collision avoidance** – The number of tags that can be read at once in a given area.

- **Readers** – Available types that support the tag.

• **Progressive use** – Perishable goods, for example, may benefit from a method of logging ambient temperature and expiration.

• **Security** – Some applications may warrant data encryption and other measures that may not be supported in all tag types.

PRINTER/ENCODERS

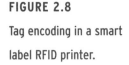 SEE PAGE 74

for encoding, printing and validation smart labels.

FIGURE 2.8

Tag encoding in a smart label RFID printer.

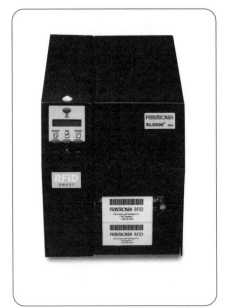

Passive WORM or EEPROM tags have no data in them. They require an encoding step to prepare them for use. Encoding can be done by a reader built into an RFID printer, or any reader that is set up for the task. Writing to a tag is more like printing a bar code than it is like reading a tag, even though both are done by an RFID reader (Fig. 2.8). That is why smart label printer makes an ideal platform for the tag encoding task.

Table 2.3 contrasts writing to a tag from reading it. When writing data to a tag, a reader has to address a tag individually. The tag must be within the proximity of the reader for the time it takes to program it, which may take several hundred milliseconds. The tag must be able to draw sufficient power from the reader to enable the programming circuitry in the tag. Isolating the right tag from others around it is very important, to prevent programming the wrong tag. In the case of an RFID printer, tags are encapsulated in a roll of smart labels, and are a known distance apart from one another. Tag isolation is achieved by the design, positioning and tuning of the reader antenna within the printer chassis. The close proximity of the antenna to the tag can be used to advantage, utilizing the properties of near-field electromagnetism to inductively couple to the tag. Closed-loop data validation and

→ SEE PAGE 75

for near-field electro-magnetism.

	Read	Write
Initial state of tag	Must have data written to it.	Empty or pre-written
Process initiated by	Reader command	Reader command
Tag internal mechanism	Memory poll circuit	EEPROM burn circuit
Response rate	Hundreds per second	Single tag takes a hundred or more milliseconds
Addressing	One to many or one to one	One to one
Sequencing	All tags or listed individuals within read area	Serially and individually
Distance sensitivity	Moderate within effective read range	Extremely sensitive to effective read range
Validation	Multiple reads	Read synchronized with physical isolation of the tag.
Error recovery	Read bar code portion of label	Print overstrike on label and encode next one.

TABLE 2.3

A comparison of tag programming to tag reading.

error-recovery mechanisms are also built into an RFID printer, making it instrumental to an on-demand tag programming and application process.

READERS

RFID readers use backscatter reflection, similar to radar, to energize tags and read their response. A reader uses its antenna to send digital information encoded in amplitude modulated (AM) waveform. A receiver circuit on the tag is able to detect the modulated field, decode the information, and use its own antenna to send a weaker AM signal response.

→ SEE PAGE 80
on reading smart labels.

Because many tags may be in the presence of a reader, they must be able to receive and manage many replies at once, potentially hundreds per second. Collision avoidance algorithms are used to allow tags to be sorted and individually selected. A reader can tell some tags to wake up and others to go to sleep to suppress chatter. Once a tag is selected, the reader is able to perform a number of operations, such as reading the identification number, and writing information to the tag in some cases. The reader then proceeds through the list to gather information from all the tags.

Several types of readers exist. A number of companies make them, and more are being developed for supply chain RFID applications. Reader types include handheld, mobile mounted (forklift or cart), fixed read-only and combination reader/encoder (Figure 2.9). In a typical distribution center, a set of readers would be configured to read any set of tags passing between them. Such a configuration is called a portal. Portals may be located at receiving dock doors, packaging lines and shipping dock doors. Mobile mounted and handheld readers can be used to check tags that are not picked up through the portal, or to locate product in the DC or on trucks.

FIGURE 2.9

Typical fixed station and handheld readers.

Photo courtesy of Alien Technology, Inc.

Photo courtesy of Applied Wireless Identification Group, Inc.

Considerations for selecting readers include:

• **Operating frequency** – Matched to tag requirements.

• **Multi-protocol** – A desirable characteristic if a variety of tags are to be read which may have different air interface protocols.

• **Meets local regulations** – Power output will be different in the USA and Europe. Frequency hopping is required in the USA and duty cycle in Europe.

• **Networking capability** – Ability to network readers together, and communicate with host computers through common interfaces (cable, twisted pair or wireless), using RS-485, TCP/IP, Ethernet, or 802.1.

• **Configurable and upgradeable** – Through network connection and firmware.

⊙ SEE PAGE 133
on reader antenna
placement.

• **Antenna** – Adapts to various conditions using dynamic auto-tuning. Can accept multiple antennas for various applications.

• **Control interfaces** – Digital input/output and control circuits for synchronization with other components on an automated line.

READER ANTENNAS

Reader antennas are the most sensitive component of an RFID system. Typical antenna set-ups for packaging line and dock door reading are shown in Figure 2.10a and b. Most reader antennas are housed in enclosures that are easy to mount, and tend to look like plain, shallow boxes. Varying the reader antenna placement is usually the easiest adjustment to make when troubleshooting a system, and one of the trickiest things to do well. The reader antenna must be placed in a position where powering the tag and receiving

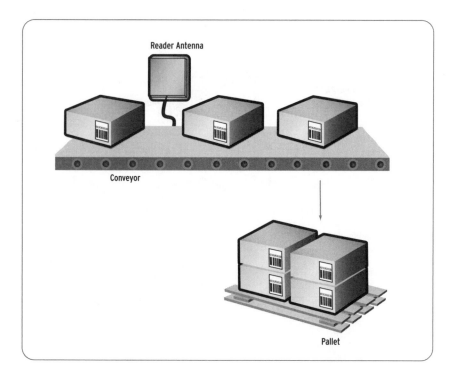

Reader Antenna

Conveyor

Pallet

FIGURE 2.10a

Antenna configuration for conveyor "curtain".

data can be optimized. Since government regulations limit the broadcast power of a reader, antenna placement is vital to achieving a high read rate.

Three characteristics of antennas contribute to tag readability:

- **Pattern or footprint** – The three-dimensional energy field created by the antenna. This is also known as the reading area.

- **Power and attenuation** – The maximum power of a reader antenna is fixed in order to meet FCC and other regulatory requirements. The signal can be decreased or attenuated, how-

FIGURE 2.10b

Antenna configuration for a dock door "portal".

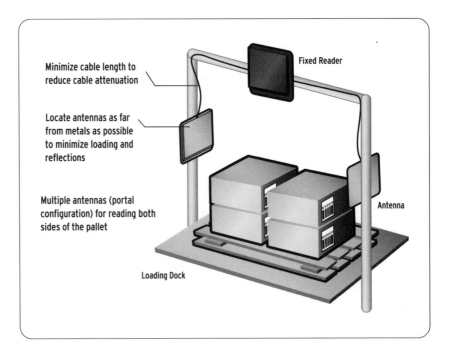

Minimize cable length to reduce cable attenuation

Locate antennas as far from metals as possible to minimize loading and reflections

Multiple antennas (portal configuration) for reading both sides of the pallet

Fixed Reader

Antenna

Loading Dock

ever, to limit the tag read window or aim it only at tags you want to read.

- **Polarization** – The orientation of the transmitted electromagnetic field.

Linear antennas typically provide the longest range, but are sensitive to the orientation of a tag (Fig. 2.11). They can be used in instances such as an RF reading curtain mounted over a conveyor. The tags would be affixed to packages in a consistent orientation to maximize readability.

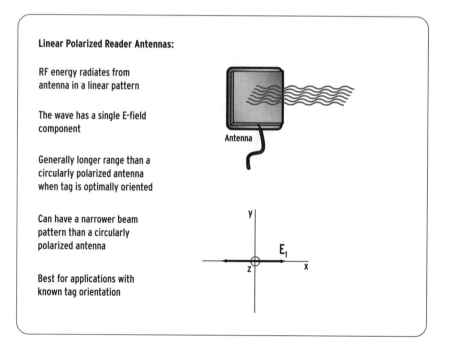

Linear Polarized Reader Antennas:

RF energy radiates from antenna in a linear pattern

The wave has a single E-field component

Antenna

Generally longer range than a circularly polarized antenna when tag is optimally oriented

Can have a narrower beam pattern than a circularly polarized antenna

Best for applications with known tag orientation

FIGURE 2.11

Linear polarization.

Circular polarization is created by an antenna designed to radiate RF energy in many directions simultaneously (Fig. 2.12). The antenna offers greater tolerance to various tag orientations, and a better ability to bounce off of and bypass obstructions. These abilities come somewhat at the expense of range and focus.

UHF antennas are nearly always externally mounted and connected to a reader via shielded and impedance matched coaxial cable. One or more antennas can be connected to a signal reader, depending on your application requirements. An antenna is first selected for the operating frequency and application (omni-directional,

FIGURE 2.12

Circular polarization.

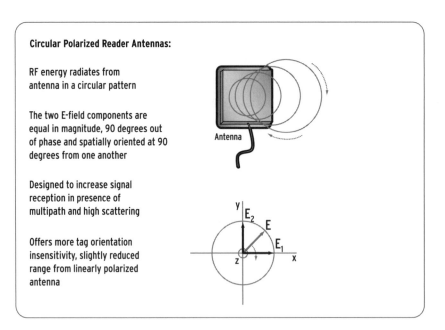

Circular Polarized Reader Antennas:

RF energy radiates from antenna in a circular pattern

The two E-field components are equal in magnitude, 90 degrees out of phase and spatially oriented at 90 degrees from one another

Designed to increase signal reception in presence of multipath and high scattering

Offers more tag orientation insensitivity, slightly reduced range from linearly polarized antenna directional, etc.). De-tuning, or signal weakening can occur due to the following:

• RF variations
• Skin-effects
• Losses due to metal proximity
• Antenna cabling losses
• Signal fading
• Proximity of other reader antennas
• Environmental variations
• Harmonic effects
• Interference from other RF sources
• Eddy Fields
• Signal reflections
• Cross talk

Some of these effects can be compensated through dynamic auto-tuning, circuits in the reader, which work with feedback from an antenna s resonance tuning parameters. In most cases, antenna placement is not pure science, and on-site adjustments are required to achieve optimal read rates.

Sources and Further Reference

Computer Networks, Third Edition, Andrew S. Tanenbaum, Prentice Hall, 1996.

A Basic Introduction to RFID Technology and Its Use in the Supply Chain, white paper, Steve Lewis, Laran RFID, January 2004.

RFID Primer, white paper, Alien Technology, 2002.

RFID for the Supply Chain: Just the Basics, white paper, Printronix, February 2004.

Understanding the Wal-Mart Initiative, white paper by John Rommel, Matrics Inc., 2004.

CHAPTER 3

From UPC to EPC

Bar codes have been with us for over fifty years. The first bar code patent was granted in 1952. It has only been in the last 30 years, however, that bar codes became ubiquitous. Mass adoption followed three converging mechanisms – industry mandates to use them, standardization, and refinements of labeling and scanning technology.

Like bar codes were 30 years ago, RFID is in very early adoption. See Figure 3.1. Unlike bar codes, RFID technology and standardization has not matured in advance of retail mandates for adoption. A dozen years went by between when UPC bar codes were standardized and retail giants mandated their use.

At this point we cannot say whether some of the more visionary proclamations about RFID (clothing tags that tell the laundry machine how to wash them, for example) will ever happen, or if RFID will ever completely replace other forms of machine identification. Mandates, standardization and technology refinements can set us on a path; it's anybody's guess where that path will lead.

RFID STANDARDS FOR THE RETAIL SUPPLY CHAIN

Standardization gained focus through the Department of Mechanical Engineering at MIT, with the formation of the Auto-ID

Center in October 1999. The Auto-ID center championed passive RF tag designs and manufacturing techniques to drive down costs. In order to keep tags as simple as possible, memory capacity was limited to a few thousand characters. An information architecture was drawn up, where tags serve as keys or pointers, wirelessly linking items to information stored in databases accessible through the internet.

→ SEE PAGE 167 on global exchange of supply chain data.

This led to the idea of the Electronic Product Code, or EPC, a comprehensive key linking an item, case or pallet to detailed information anywhere in the supply chain. The goal however, was not to replace bar codes. Rather, the goal was to create a migration

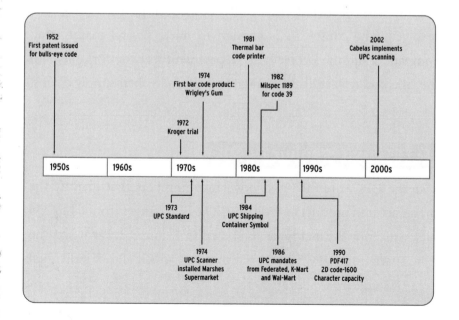

FIGURE 3.1

Bar codes gained mass adoption over a 30-year period.

path for companies to move from bar code to RFID.

The Auto-ID Center officially closed October 26th, 2003. It passed its work on to a new worldwide standards organization, called EPCglobal. EPCglobal is a joint venture of EAN International and the Uniform Code Council, which administers the UPC bar code. Together, EAN International and the UCC represent 100 member organizations worldwide with more than one million members representing 102 countries.

⊙ SEE PAGE 89

on industry initiatives.

Table 3.1 lists some of the standards organization involved in one way or another with RFID. Although differences exist among these organizations, a breathtaking amount of consensus and standards making has occurred, especially considering the short time involved. The consensus is enough to make several major retail companies and the US Defense Department feel comfortable with the risks and rewards, and mandate adoption by their supply chains.

ELECTRONIC PRODUCT CODE

Like the UPC, the EPC is divided into numbers that identify the manufacturer, product and version. The EPC ranges from 64 to 256 bits, with four distinct fields as shown in Figure 3.2. What sets the EPC apart from UPC is its serial number, which allows individual item tracking.

⊙ SEE FIG. 3.2

EPC Format

The EPC format is an open format capable of describing physical entities for a number of purposes, including supply chain RFID tag applications. The format can also be used in bar coding and other machine-readable encoding applications. The general format for EPC tag data includes these sections:

Organization	Mission	Website	Domain
ANSI	Commercial trade standards	www.ansi.org	USA
APEC	Commercial trade standards	www.apec.org	Asia-Pacific
APICS	Supply chain business process	www.apics.org	Global
Auto ID Center	RFID standards	www.epcglobalinc.org	Replaced by EPCglobal
CEN	Commercial trade standards	www.cenorm.be	Europe
CEPT	RF broadcast standards	www.cept.org	Europe
EAN	Supply chain ID & EDI	www.ean-int.org	Global
ECCC	Bar codes and e-commerce	www.eccc.org	Canada
EPCglobal	EPC & RFID	www.epcglobalinc.org	Global
FCC	RF broadcast regulations	www.fcc.gov	USA
ISO	Commercial trade standards	www.iso.ch	Global
MPHPT	RF broadcast standards	www.soumu.go.jp	Japan
UCC	Supply chain & EDI	www.uc-council.org	USA

TABLE 3.1

Standards organizations involved in RFID.

Header – The 8-bit header identifies the version number of the code itself.

EPC Manager – Identifies an organizational entity (e.g., a company, a city government) that is responsible for maintaining the numbers in subsequent fields – Object Class and Serial Number. EPCglobal assigns the General Manager Number to an entity, and ensures that each General Manager Number is unique.

Object Class – Refers to the exact type of product, similar to a SKU (stock keeping unit) The object class is used by an EPC managing entity to identify trade items. These object class numbers, of course, must be unique within each General Manager Number domain. Examples of Object Classes could include SKUs of consumer-packaged goods, or road signs, lighting poles or other highway structures where the managing entity is a county.

Serial Number – A unique identifier for the item within each object class. The managing entity is responsible for assigning

FIGURE 3.2

Electronic Product
Code format.

ELECTRONIC PRODUCT CODE TYPE 1 96-BIT

02 . 0000A68 . 00010D . 00112DED

Header	EPC Manager	Object Class	Serial Number
8-bits	28-bits	24-bits	36-bits

unique, non-repeating serial numbers for every instance within each object class.

The EPC Type 1 number, at 96 bits in length, will accommodate as many as 268 million companies, each having 16 million classes, with 68 billion serial numbers in each class. In Class 1 tags, an additional 32 bits of the EPC are for unique item information (item description, ultimate destination, special handling instructions, etc.) that can be reused at any point in the supply chain.

⚪(!)
The 96 bit EPC is more than enough to serialize every product world wide for years to come.

EPC Representations of Standard Identity Types

EPC versions of six identity types are described in the EAN.UCC tag standard working draft (Fig. 3.3). These identity types were developed as bar code standards, and the tag standard proposes EPC coding schemes for them.

Global Trade Identity Number (GTIN) – This is the formal term for an EPC used to identify items that are products and services traded by companies. To create a unique identifier for individual items, the GTIN is augmented with a serial number, which the managing entity is responsible for assigning uniquely to individual object classes. The combination of GTIN and a unique serial number is called a Serialized GTIN (SGTIN).

Serial Shipping Container Code (SSCC) – The SSCC is assigned uniquely by the managing entity to a specific shipping unit. A section of the code is set to quickly distinguish basic logistic types such as a pallet from a case.

Global Location Number (GLN) – GLN can represent functional, physical or legal entities. Examples of functional entities are a purchasing or accounts payable department. Physical entities might include a particular warehouse room or loading dock. Legal entities can include whole companies, subsidiaries and divisions.

Global Returnable Asset Identifier (GRAI) – Like a GTIN, except the object class is used to identify the asset type. Examples of

FIGURE 3.3

Standard Identity types and what they represent.

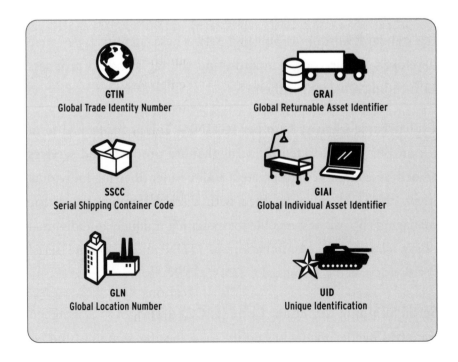

GTIN
Global Trade Identity Number

GRAI
Global Returnable Asset Identifier

SSCC
Serial Shipping Container Code

GIAI
Global Individual Asset Identifier

GLN
Global Location Number

UID
Unique Identification

returnable assets include barrels, pallets, gas cylinders, beer kegs, rail cars and trailers.

Global Individual Asset Identifier (GIAI) – GIAI is used by a company to label fixed inventory or any property used to carry on the business of the company. Examples are hospital beds, computers and delivery vehicles.

Unique Identification (UID) – UID is the DoD asset tracking number, which is being harmonized into the EPC standard.

EVOLUTION OF TAG CLASSES

RFID tags have strayed somewhat from the neat classifications originally conceived by the Auto-ID Center. As EPC evolves as a classification system, EPCglobal has moved toward definitions of tags that combine both the data and physical layer. The standards evolution is shown in Figure 3.4.

Class 0 – These tags are passive, UHF based and are factory programmed. They are the simplest type of tags. Because the ID numbers in the tags are preset, they are associated arbitrarily to cases and pallets through a host computer at the packaging stage. Class 0 tags are well suited to applications such as anti-theft devices, where their presence is detected and the ID number is of little consequence. Wal-Mart has approved a 96-bit Class 0 version for case/pallet RFID. A Class 0+ version is a WORM version to support EPC package origination encoding.

Class 1 – Class 1 tags are passive, UHF or HF (13.56MHz) based, have a WORM structure and are field programmable. Class 0 and Class 1 tags, which tend to be used in similar supply chain applications, are not currently interoperable. The next generation that will replace Class 0 and Class 1 tags is UHF Generation 2 Foundation Protocol, which will be a full read/write 128-bit RFID tag; 96 bits will be for the EPC and an additional 32 bits will be for error correction and a kill command. Because UHF Gen 2 is likely to be a hybrid of the existing standard frameworks, it marks the start of a unified passive RFID standard for supply chain applications. The UHF Gen 2 standard is expected to be finalized late-2004.

FIGURE 3.4

Passive tags are evolving toward a common standard.

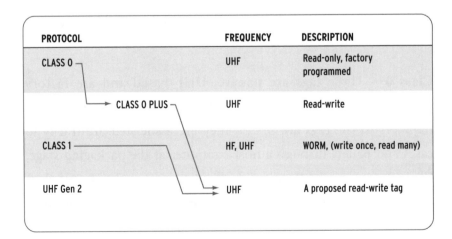

PROTOCOL	FREQUENCY	DESCRIPTION
CLASS 0	UHF	Read-only, factory programmed
CLASS 0 PLUS	UHF	Read-write
CLASS 1	HF, UHF	WORM, (write once, read many)
UHF Gen 2	UHF	A proposed read-write tag

Class 3 tags are a passive/active hybrid having battery backup, which acts as an internal power source, but it remains in a passive mode until activated by a reader. Class 4 tags are active tags. Each of these class standards has not been fully ratified, and in many cases is evolving rapidly towards new standard versions.

EPC AND RFID COMPARED TO UPC AND BAR CODES

For the past 25 years bar codes have been the primary means of identifying products in the supply chain. Bar codes have been effective, but have limitations. The key attributes to consider when comparing RFID and bar coding center around reading capability, reading speed, tag or label durability, amount of information, flexibility of information, cost and standards. A migration toward RFID involves a number of considerations, one of which is whether it should complement or replace bar coding for the application.

Read Method – Seeing and hearing are two different things as they say (Fig. 3.5). Line of sight has its advantages. Bar code optical readers offer an absolute visual verification. The reader signals a good read within its aim, and a bad read is immediately associated with a specific label and item. This is a one-to-one relationship. RFID is more like hearing. It does not require line-of-sight to read the tag information. The radio frequency (RF) signal is capable of traveling through most materials. This is particularly advanta-

→ SEE FIG. 3.5

⊕ **SEE PAGE 80**

on reading smart labels.

geous in warehouse receiving operations, and in operations where information needs to be collected from items that may have an inconsistent orientation, such as distribution center sorting applications. An RFID reader is able to distinguish and interface with an individual tag despite multiple tags that may be within the given read range. The discrimination of tags, however, does not include with it the absolute physical location identification that a bar code reader has when it is aimed at a specific point on a packaging line. Tags that do not respond (quiet tags) for one reason or another require a manual search and verify step, or diversion of the entire pallet for further investigation. Suddenly, a whole

FIGURE 3.5

Roughly speaking, a bar code is like seeing and RFID like hearing.

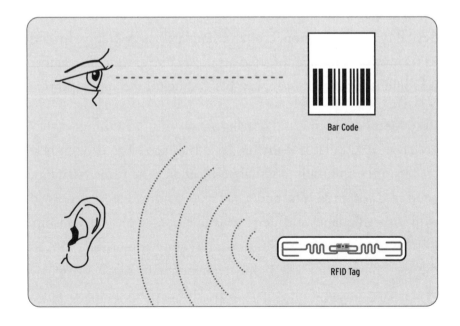

Bar Code

RFID Tag

pallet is held up. A batch process now requires batch level recovery mechanisms.

Read Speed – RFID tags can be read far more rapidly than a bar code label, at theoretical rates up to 500 per second (Fig. 3.6). This far surpasses the one-at-a-time reading speed of bar code. The speed of RFID has great value in high-volume receiving and shipping applications where a large number of items need to be accounted for quickly. For example, when receiving a pallet of tagged cases in a warehouse, an RFID reader can potentially identify all cases without having to break the pallet down and scan each individually.

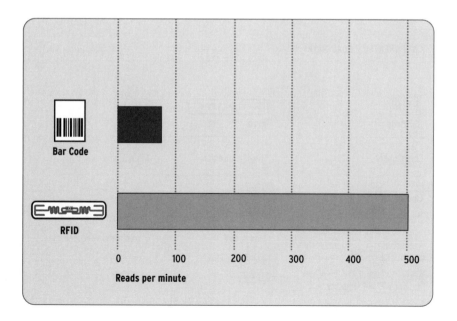

FIGURE 3.6

Comparison of reading speeds of a bar code versus RFID.

Readability – Read rates approaching 100% are possible with bar codes in high-speed automated lines engineered for bar code identification. The engineering practice is established and results are repeatable. RFID offers the promise of perhaps better read rates at equal or greater line speed, but the engineering practice is in its early stage. Achieving a high accuracy goal, such as Wal-Mart's 100% mandate for RFID, may be non-trivial for many companies. Figure 3.7 summarizes some of the components of read accuracy.

→ SEE PAGE 135

on characteristics affecting read rate.

Durability – RFID tags can be encased in hardened plastic substrates or other materials. Although they are significantly more durable than paper bar code labels, both depend on adhesive to hold them intact and attached to an item. Bar codes etched on

FIGURE 3.7

100% readability will involve process, environmental and technical engineering.

COMPONENTS OF READ-RATE:

Bar Code	RFID	
Environment	Reader Antenna	Tag Antenna
Line speed	Environment	Case analysis
Label to reader orientation and distance	Line speed	Encoded and validated
Meets symbology specifications for contrast and legibility		

metal or plastic have proven reasonably durable over the years. RFID has been used to track engine blocks throughout their production process, which is often too harsh for a bar code. Further, the durable nature of RFID tags allows them to last longer than bar code labels. The Achilles' heal of an RFID tag is the mating point of the antenna to the chip. A cut that severs through the mating point will disable the tag, whereas as a bar code may be only slightly degraded.

Data Storage – UPC identifies an item classification, but EPC can identify an individual item through an assigned serial number. The traditional linear or 1D bar code stores up to 100 characters, and a 2D bar code can store 1,000 characters or more. See Figure 3.8. High-end RFID tags may contain several kilobytes of memory (several thousand characters). This increased information storage capability creates a portable database of information, allowing a greater number of product attributes to be tracked, such as date of manufacture, time spent in transit, location of distribution center holding the item, expiration date or last date of service. On the other hand, a typical bar code shipping label contains an enormous amount of information in human readable and bar code form, including addresses for shipping, content markings, means of conveyance, etc. A smart label combines both storage capabilities.

⊙ SEE FIG. 3.8

Flexibility of Information – With respect to information dynamics,

→ SEE PAGE 161

on smart labeling systems integration.

RFID tags are able to support read/write operations, enabling real-time information updates as an item moves throughout the supply chain. The total information model for passive tag RFID, however, does not depend on re-writing the tag. The EPC number written on a tag serves as a key to information stored in external databases. These databases support the concept of recording the history of a tagged item throughout its lifecycle. See Figure 3.9. This feature can be of critical importance as production schedules, delivery dates and locations, and shipment contents can change on a regular basis. Bar codes, of course, can also be used to record an EPC number, and serve as keys to external databases. In most supply chain applications, however, they are not used this way.

FIGURE 3.8

Comparison of data storage capacities.

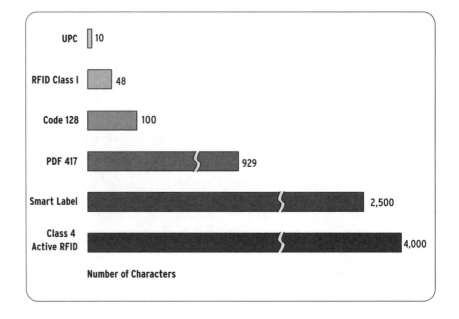

Information Redundancy – RFID tags hold information in captive form, offering it only through a reader tuned to accept it. System integrity is non-linear – you either accept or reject what the reader says. Bar codes, on the other hand, usually have a human-readable character format adjacent to the bar (Fig. 3.10). This permits a direct recovery should a bar code fail to read. Think of how many times you have avoided a long wait at the grocery checkout counter as the clerk picks up the package and keys in the bar code number after it failed to scan. Smart labels, containing bar code and human readable characters along with the RFID tag, may offer the best combination of information redundancy and integrity.

(→) SEE FIG. 3.10

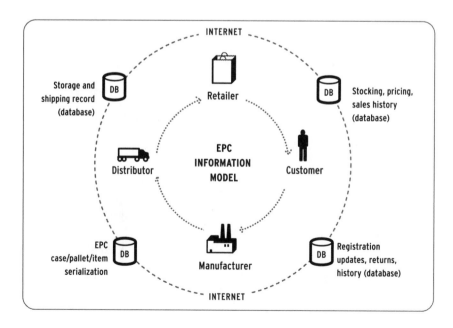

FIGURE 3.9
RF tags have potential over the entire product life cycle.

Security – Both RFID tags and bar codes have encryption routines that support various information security requirements. Two-dimensional matrix bar codes are impossible for a human to read, just as with RFID tags. Some tags, however, support a password scheme that can render them unreadable to reading systems that do not use the password to access the EPC code.

Cost – RFID requires new capital and operating investments. Figure 3.11 shows the typical investments and costs. A return on investment (ROI) cost calculation may justify RFID only when the risk of not doing

FIGURE 3.10

Comparison of information redundancy.

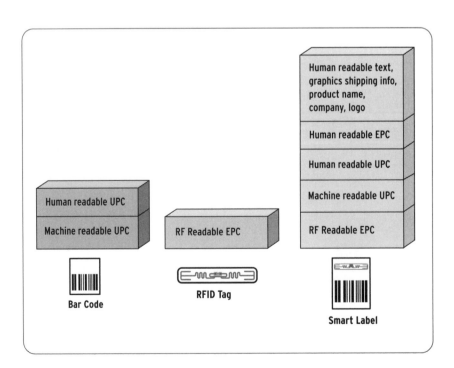

so means losing a prime customer, such as Wal-Mart, Target or DoD. A pragmatic approach, where the investment level keeps you in business as you gain knowledge and experience, may be the least cost approach.

ADDRESSING CONSUMER CONCERNS ABOUT RFID

Although RFID implementations are currently focused on supply chain case and pallet tracking, the long-term goals of the retail industry include the tagging of items. Early trials and media stories about them have raised public concerns about RFID, and public advocacy against its use. The concerns relate to the privacy of individuals as they acquire and use products with embedded RFID tags.

The retail industry realizes that public acceptance cannot be taken for granted. In response, EPCglobal has adopted guidelines for the

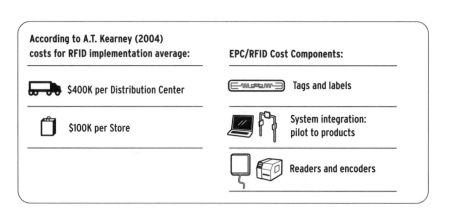

According to A.T. Kearney (2004) costs for RFID implementation average:

$400K per Distribution Center

$100K per Store

EPC/RFID Cost Components:

Tags and labels

System integration: pilot to products

Readers and encoders

FIGURE 3.11

Typical startup costs for a retail distribution center RFID implementation.

use of EPC on consumer products. These guidelines are certain to evolve, and currently address these areas of concern:

Consumer notice – Products and packaging must be clearly marked when they have an RFID tag. The EPC logo is the suggested marking Smart labels serve to clearly indicate the presence of an RFID tag by having them preprinted with identifying information.

Consumer choice – Although this is not anticipated to be as big an issue at the case level, there needs to be a way for consumers to disable or remove the tag. A smart label can be peeled off a carton, so it passes that test. Embedded tags will have to be marked as such, with instructions on how to extract or disable them should the consumer wish to retain the carton without the tag. Tag manufacturers are including a kill command in the chip, which could be used at some point to render the tag inoperable.

Consumer education – The responsibility for consumer education about RFID, the EPC logo and where to obtain more information will have to be shared amongst members of EPC global.

For instance, informational labeling on packaging that is used at the retail level will have to be agreed upon and used.

Record use, retention and security – Policies regarding the types of information collected and retained must reflect state, national and regional regulations. Companies will have to make those policies public, on their web site and by other means, and their actual records and practices may be subject to audits and legal oversight.

Sources and Further Reference:

The Bar Code Book, by Roger C. Palmer, Helmers Publishing, Peterborough NH, 1989.

RFID Explained, A Basic Overview, white paper, Robert W. Baird & Company, February 2004.

A Basic Introduction to RFID Technology and Its Use in the Supply Chain, white paper, by Steve Lewis, Laran RFID, January 2004.

Guidelines on EPC for Consumer Products, available: www.epcglobalinc.org.

CHAPTER 4

From Bar Codes to Smart Labels

IN THIS CHAPTER:

How smart labels combine RFID with bar code for case/pallet pilot applications.

Smart labels are shipping labels with embedded RFID tags. They offer promise in helping organizations deploy RFID for compliance with retail industry and DoD mandates. Smart labels allow you to retain bar code/shipping label information in the same or similar format to what you are currently using, while adding RFID. Smart labels provide these general benefits for supply chain applications:

- A convenient and economical way to package RFID tags and stream them into the distribution process
- A richer data set than either a tag or barcode by themselves
- Back up and redundancy, by having bar code and human-readable text together with electronic data, should one method of identification fail
- Complies with consumer and industry guidelines for having a visible indication that a package has a RFID tag
- Can be produced on-demand, or pre-printed and pre-coded for batch processing
- Labels provide added protection to the tag from heat, dust and humidity

→ SEE PAGE 143

on smart labeling approaches.

Smart labels may be the easiest, least disruptive, least cost way to implement RFID in your facility. They are not only appropriate for case and pallet supply chain applications, they may also be used in a number of applications "within the 4 walls," including receiving, routing, stocking, work-in-process, HAZMAT and asset handling.

If you are already producing bar code labels, a migration to smart labels could involve the integration and re-use of an established process, where:

- On-demand printing and application flexibility is maintained
- Labeling is done at appropriate points in the packaging/shipping process
- RFID integration fits within the small footprint of a smart label printer
- Both automated and operator-assisted application methods are available
- Tag encoding is done predictably and reliably, without custom engineering
- Validation and error recovery is built into the system
- Encoding and printing commands share an established host computer to shop floor network
- System migration and integration can be simplified using conversion tools and software modules from multiple middleware and supply chain execution system suppliers, so you don't have to rewrite applications

ANATOMY OF A SMART LABEL

Figure 4.1 provides two views of a smart label. The surface area is used for standard bar code and label text. The RFID tag is sandwiched in the middle. The label sandwich consists of six parts – the liner or carrier sheet, the liner release coating, the tag inlay, an adhesive, the label material and a label topcoat.

Labels are printed using a thermal print process. A print head, embedded with a matrix of tiny wires that are precisely controlled, uses heat to transfer a wax or resin ink to the label. The ink is on a separate ribbon inside the printer. Smart labels come in rolls of

FIGURE 4.1

Two views of a smart label.

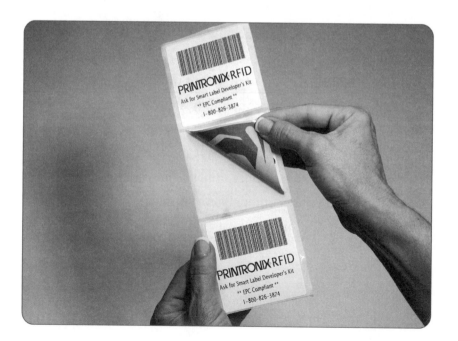

various sizes which, along with the ribbon roll, are mounted inside the printer/encoder.

SELECTING THE RIGHT SMART LABEL FOR THE JOB

Smart labels are available in bulk roll quantities in a variety and sizes and types. You'll find formats that allow smart labels to substitute for existing bar code labels designed to meet corporate and supply chain standards (mil spec labels, freezer grade labels, pharmaceutical chain of custody labels, etc.). Three basic criteria should drive selection of the right label:

Labeling requirement – The information on the surface of the label and embedded in the tag needs to identify the package contents and its status in the supply chain. Supply chain status could include originator, shipper, ship to and any special handling requirements. Figure 4.2 shows a typical label format, with address areas and machine readable codes. The amount of information on the front of this label exceeds what is stored in EPC format on the RFID tag.

⊙ SEE FIG. 4.2

Application requirement – Tag readability is a major factor. A complete case analysis is required to ensure tag readability. Case analysis should include a comprehensive assessment of package contents, packaging design, label placement and the package labeling process. Package contents and packaging design may affect tag

⊙ SEE PAGE 127
on case analysis.

FIGURE 4.2 Typical format for case labeling.

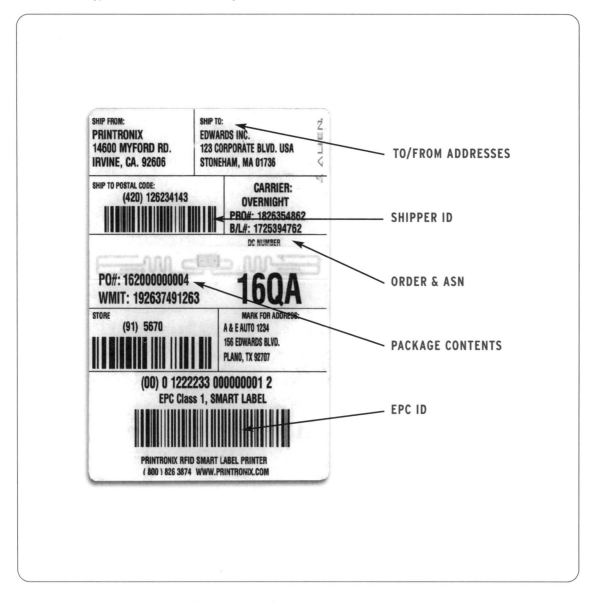

readability, particularly if metals, liquids, high carbon or salt content is involved. This affects label placement on the package, and package orientation with respect to readers as it moves through the process. Storage and handling requirements determine whether you need such things as freezer-grade adhesive, or polyester rather than paper label material to withstand heat.

Compliance requirement – Customer specifications may dictate everything from the size of label used, the information printed on label surface, where the label should appear on a package, the EPC information in the tag, the tag/reader air protocol, and acceptance criteria. Acceptance criteria may include such things as read rate (100%), advanced shipping notice (ASN) integration, validations, record keeping and problem handling (charge-backs). The stringency of your customer's acceptance criteria will greatly affect the design criticality of your process.

→ SEE PAGE 89
on industry initiatives.

→ SEE PAGE 93
on ASN integration.

Clearly, putting an RFID tag or smart label on a product is not as easy as applying a bar code label. The engineering trial and error involved is one of the biggest challenges facing companies who have to meet a RFID compliance mandate. Application requirements, as defined by a case analysis, usually determine the choice of smart label. Case analysis will match the application to a specific tag and antenna design. Figure 4.3 shows several tag antenna designs available on smart labels.

→ SEE FIG. 4.3

FIGURE 4.3

Smart labels with tags for various uses.

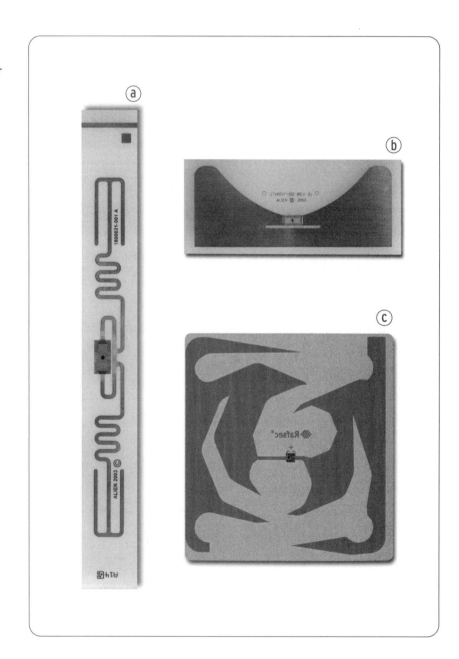

Tag and label manufacturers are greatly expanding their offerings and designs to meet various application requirements. The "squiggle-tag" (Fig. 4.3a) is designed for general purpose applications. The "U-tag" (Fig. 4.3b) is designed to electrically couple with a detergent bottle to enhance its reception. The "psychedelic-tag" (Fig. 4.3c) is designed to function even when packages are rotated in various orientations.

LABEL CERTIFICATION

Although smart labels are available from a number of sources, certain labels may be incompatible with certain printer/encoders. Certified labels have been pre-tested to eliminate incompatibility, and ensure optimum performance. A one percent error rate on labels, for example, when production throughput is 40 labels/minute, would result in almost 200 rejects a day.

Reasons for incompatibility include:
- Tag type needs to be matched to the encoder.
- Tag position needs to be at a position within the label so that it is correctly oriented to the encoder head.
- Label surface material needs to be suitable for high quality production thermal printing.
- Adhesive needs to be matched to the package surface to which it will be applied.

- Labels (paper versus synthetic) may be mismatched to the ribbon stock (wax versus resin).

In addition, retail package labeling requires a high quality label matched to its purpose. Labels that are difficult to peel, or do not consistently peel from their backing because of poor die cuts will fail and require operator intervention. Poor adhesive application during the label converting process can result in labels lifting as they travel through a printer, causing a jam. The core diameter of label rolls need to be matched to label sizes to reduce the amount of induced curl on labels as the roll unwinds to its core. Curl or adhesion problems will affect how well the vacuum system on an applicator can hold a label in place. Label carrier sheet stiffness can also contribute to applicator problems by making it difficult for the label to separate as it passes through the peel bar.

Label certification is especially important with RFID, since the programming of the tag and synchronization of tag data with printed label data occurs at the printer, all within a second's time.

ENCODING, PRINTING AND VALIDATING SMART LABELS

Initially, passive UHF tags have no data in them. They require an encoding step to load data into them. Encoding can be done by a reader built into an RFID printer, or any reader that is set up for the task. Writing to a tag is more like printing a bar code than reading a tag, even though both reading and writing can be done by an RFID reader. A smart label printer makes an ideal platform for the tag encoding task for a number of reasons:

Isolation – When reading tags, the reader starts by compiling a list of tags, which it can poll individually if it wants. When writing data to a tag, a reader has to address a tag individually. Isolating the right tag from others around it is very important, to prevent programming the wrong tag. Blank tags won't respond to a call. Some tags have null data in them, inserted by the tag manufacturer during parameter testing. A reader calling to those tags may get the same response from all of them. The only way to synchronize the reader with a blank or null-data tag is by uniquely positioning the tag within a precise read window. That way, the reader has a tag, one tag only, and the right tag within its grasp.

Proximity – Within one wavelength from a power transmission source, conductive materials experience the affects of near-field electromagnetism. See Figure 4.4. Within the near-field, magnetic coupling occurs. Power flows in the direction of the magnetic field. Near-field energy transfer is more like that of an electric motor or a power transformer. At greater than one wavelength, radio waves separate and propagate away, no longer aligned with the magnetic field. Energy drops off by the square of the distance. At 900 Hz, the

⊙ SEE FIG. 4.4

near-field breaks down within a few inches of the source, therefore a reader antenna more than a foot away from a tag cannot take advantage of near-field magnetic coupling.

Power and duration – Compared to a read command, a write command requires a higher power level and longer duration. The tag must be able to draw sufficient power from the reader to drive the programming circuitry in the tag. The tag must be within the proximity of the reader for the entire time it takes to program it.

You can see that the challenge isn't so much how as it is which tag to program. In the case of an RFID printer, tags are encapsulated in a roll of smart labels, and are a known distance apart from one another. Tag isolation is achieved by the design, positioning and tuning of the

FIGURE 4.4

Near-field and far field electomagnetism.

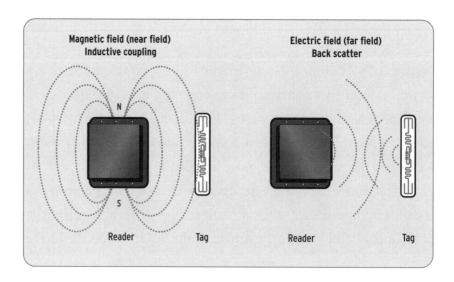

Magnetic field (near field)
Inductive coupling

Electric field (far field)
Back scatter

N

S

Reader   Tag   Reader   Tag

encoder antenna within the printer chassis. The close proximity of the antenna to the tag is used to advantage, by utilizing the properties of near-field electromagnetism to inductively couple the tag. Because of the encoding time needed, label production time is somewhat less than it takes to produce a bar code label only. The time trade-off, however, is more than compensated for by precise process control, high duty cycles, and validation and error recovery routines that eliminate bad tags and labels being applied to a package.

→ **SEE PAGE 143**
for smart labeling approaches.

Figure 4.5 shows the interior side view of a printer/encoder. The sequence of operation is as follows:

→ **SEE FIG. 4.5**

1. The printer receives its commands from the host computer. Once a label is in position, the reader does a pre-check. Some tags have null data encoded by the tag manufacturer as a part of its parameter testing system. The null data helps distinguish good tags from quiet tags that may have escaped detection during the smart label converting process. If a tag is quiet during the pre-check, the printer ejects it by marking it with a strike-over (Fig. 4.6).

→ **SEE FIG. 4.6**

2. A single command by the reader programs the tag. The command includes an erase, write, read back and verify sequence, along with the specific EPC data to be written to the tag. Tags cannot be partially written to. A write command replaces data that might be in the tag. If the read back and verify is not successful, the command is repeated.

FROM BAR CODES TO SMART LABELS:
Encoding, Printing and Validating Smart Labels

FIGURE 4.5 Side view of the inside of a smart label printer.

3. The reader then performs an explicit read back, to again match the tag data reply with what it was expecting to hear.

4. The thermal head prints the label, using the bar code and text character job stream associated with the EPC number.

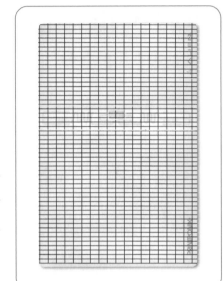

FIGURE 4.6
Smart label with strike over pattern to indicate a bad label.

5. If the printer has ODV (Online Date Validation) installed, it also reads the bar code information with a bar code scanner mounted in-line with the label exit path. ODV compares the bar code with AIM standards for dimensional tolerance, edge roughness, spots, voids, reflectance, quiet zones and encodation. If the label does not pass, the printer reverses its path, overstrikes the label to indicate that it failed the test, and ejects the label.

6. A record of the production sequence is sent to the host computer.

7. If the printer is integrated with an applicator or other packaging line components, it will communicate with logic controllers to sequence next steps in the production line.

⊙ **SEE PAGE 143**
for smart label print & apply approaches.

READING SMART LABELS

Smart labels are a ready-fit format for most RFID implementations.

FIGURE 4.7

Operational data flow.

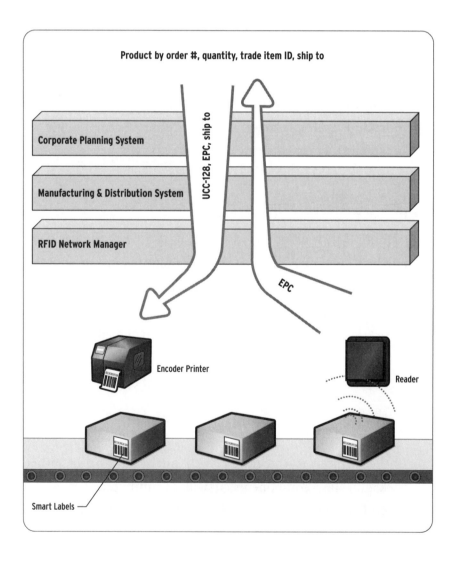

EPC data flows down to them through a host computer system and printer/encoder. See Figure 4.7. Labels are applied to parts, products, packages and pallets, uniquely identifying them. Smart-labeled objects are now linked up by radio frequency to a supply chain execution system. To help complete the picture, lets look at how they are read. Keep in mind, RFID tags must first be programmed, or written to, before they can be used. They start out as either blanks or with null data embedded by the tag manufacturer. Readers have a number of commands they can use to communicate.

A version 1 tag command format contains a unique identifier, error detection/correction code applied to that identifier, and a short password, as shown in Figure 4.8. The unique identifier is typically the EPC code. The error detection/correction code is a Cyclic Redundancy Check (CRC). There are no restrictions on the password.

Reader Commands

The version 1 standard describes a number of commands that a

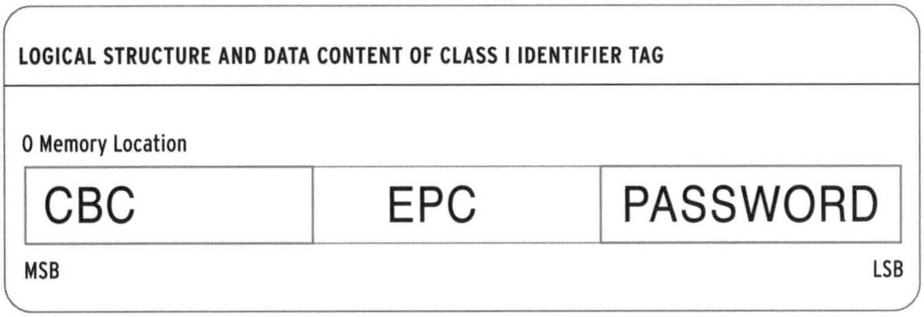

LOGICAL STRUCTURE AND DATA CONTENT OF CLASS I IDENTIFIER TAG

0 Memory Location

CBC	EPC	PASSWORD

MSB   LSB

FIGURE 4.8

Class 1 tag data structure.

reader uses to communicate with a tag. The command set is sure to change over time. Typical commands that a reader uses include:

Scroll All – All tags reply by communicating an eight (8)-bit preamble, followed by the CRC (sent MSB first), followed by their EPC.

Scroll by ID – Tags matching a specific value reply by communicating an eight (8)-bit preamble, followed by the CRC (sent MSB first), followed by their EPC.

Ping – Tags matching a specific value reply by sending eight (8)-bits of the tag identifier. The difference between Ping and Scroll is that Pinged tags do not reply with their EPC codes.

Quiet – Tags matching a specific value enter a quiet mode, where they no longer respond to or execute reader commands. This mode of operation is maintained until a proper Talk command is received and correctly interpreted or power has been removed from the tag for at least 1 second and at most 10 seconds.

Talk – Tags matching a specific value enter an active mode, where they respond to commands from the reader. This active mode of operation is the same mode that a Tag powers up into. This mode of operation is maintained until a proper Quiet command is received and correctly interpreted or power has been removed from the tag for at least 1 second and at most 10 seconds.

Kill – Tags matching the complete tag identifier, CRC and an eight (8)-bit Password are permanently deactivated and will no longer respond to or execute reader commands. This "self-destruct" command renders the tag inactive forever.

Reader Operating Modes

For most applications, readers will operate in one of two ways, either autonomously or as directed/interactive devices. The air interface protocol is the same. They transmit signals in half-duplex AM modulation, using frequency hopping across approximately 60 bands between 902-928 MHz (USA). Frequency hopping is an FCC requirement to minimize interference with other RF devices. Anti-collision algorithms, combined with a sequence of scroll,

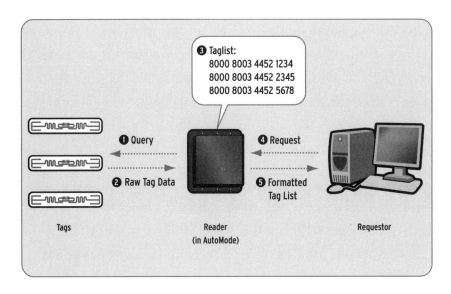

FIGURE 4.9

A reader set to read smart labels autonomously.

quiet and talk commands, are used to read and sort multiple simultaneous incoming tag signals.

Autonomous mode – A reader can be set to continuously operate, accumulating lists of tags in its memory. Tag lists represent a dynamic picture of the current tag population in its read window. See Figure 4.9. As tags respond to reader broadcasts, they are put on the list. If they don't respond they are dropped from the most recent list stored in memory list. A persist time is set to determine the duration between the time a tag was last read and when it is removed from the list. A host system on the network can receive a list of tags from the reader whenever it chooses to listen. The information available to the host would include the reader location, time read, the size of the tag list, and the IDs of the tags on the list.

Directed/interactive mode – Readers in this mode will respond to commands from the network host. The host can instruct the reader to gather a list of tags within its read window, or look for a specific tag. In both cases the reader starts by gathering a list. Once it completes the host command, the reader waits until it receives another.

SMART LABELS COMPARED TO OTHER APPROACHES

Smart labels can play a key role in an RFID migration strategy. An externally applied adhesive smart label is the easiest, quickest way to go from "no tag" to "tag selectively," paving the way to "tag everything." They bridge laboratory and pilot applications by putting tags into play in various environments, so that tag placement, tag orientation, read range, read rates, reader placement, and data management issues can be identified and resolved. In most cases they provide a production or near production line solution at least cost. In all cases they offer a backup reading capability to aid troubleshooting and recovery.

Let's look at the pros and cons of other approaches:

Boxes with built-in RFID inlays – Corrugated packaging manufacturers intend to offer boxes with RFID inlays in the coming years. The advantage for packaging lines is the elimination of a label application step. Disadvantages include lack of clear marking for RF, lack of a backup identification of what has been programmed into the tag, and a huge potential for rework costs because once the box is built, tag location can't be changed.

Pallets and totes with permanent RFID tags – Such pallets offer a simple approach to pallet level identification, especially where the tags are programmable, and the pallets are dedicated to a specific route in the supply chain. Similarly, for direct to store shipments, totes with fixed tags may gain acceptability. This approach might suit the tagging of work-in-process goods during manufacturing as well. Disadvantages when compared to smart labels include the difficulty in encoding unique EPCs with each use, and re-purposing totes and pallet loads.

RFID tags only – Another possible approach is applying adhesive backed inlays directly to cases and then programming them when they travel through the packaging line. Although this may appear to be a less-costly, simpler method, there are a number of drawbacks. Tags will most likely not be available with the broad selection of adhesives that exist today for labels. RFID tags by themselves do not provide the human readable and bar code backup that smart labels provide. They are subject to the same issues of adhesion to moist, frozen or non-flat surfaces as smart labels, but without being as obvious when they fall off. Consumer privacy groups have also requested a clear indication when an RFID tag is present, which a smart label provides. In addition, smart label printers have validation/recovery routines that prevent a bad tag from ever being applied to a case. Applying just the tag, then programming it while it's moving down the packaging line, necessities package-level rework should the tag fail to read.

Table 4.1 summarizes the advantages and disadvantages of various approaches compared to smart labels. The observations made here are in the context of pilot applications and initial use over the 2 years. In the future we can expect to see a variety of approaches to integrating RFID with today's high-speed retail packaging environments.

→ SEE PAGE 109

on pilot implementations.

Sources and Further Information

RFID for the Supply Chain: Just the Basics, white paper, Printronix, February 2004.

A Basic Introduction to RFID Technology and Its Use in the Supply Chain, white paper, Steve Lewis, Laran RFID, January 2004.

RFID Primer, white paper, Alien Technology, 2002.

8 Supply Problems when Using Label Printer Applicators, white paper, Fox IV Technologies, available: www.foxiv.com.

	Boxes with built in tags	Pallets/totes with permanent tags	Tags directly applied to cases	Smart labels
Tag placement	Fixed and pre-engineered	Fixed and pre-engineered	Flexible	Flexible
Tag application	Built-in	Built-in	Applicator	By hand or by applicator
Encoding sequence	Programmed on packaging line	Programmed at shipping dock	Applied then programmed	Programmed then applied
Activation	Curtain	Portal	Curtain	Printer/encoder device
Error recovery sequence	Case unpack & repack	Pallet unpack & repack	Rework	Detection & recovery before applied to case

TABLE 4.1

Comparison of various tagging approaches to smart labels.

CHAPTER 5

Industry Initiatives

WAL-MART

The Wal-Mart RFID journey began early in the decade with a trial involving Procter & Gamble, International Paper, Unilever, Gillette, Johnson & Johnson and Kraft. Scientists from the Auto ID Center and the Uniform Code Council lent support to the three-phase trial. Phase I, which commenced in late 2001, tested small numbers of products and volumes at the pallet level. Phase II began in early 2002 and tested higher case level volumes and a greater mix. Phase III was designed to test individual items tracking down to the unit level, but was scrubbed in 2003 in favor of other Wal-Mart initiatives.

In June of 2003, Wal-Mart announced to its top 100 suppliers that they would be required to adopt RFID technology by 2005. Wal-Mart met with its top 100 suppliers, along with 37 other suppliers who volunteered, to lay out the specifics of its RFID mandate. The mandate requires suppliers to use RFID tags on their pallets and cases shipped to three Wal-Mart distribution centers in Texas by January 1, 2005. Additional suppliers, DC's, Wal-Mart and SAM'S CLUB stores will be phased in during 2005 and 2006. Figure 5.1 shows the implementation timeline.

Why is Wal-Mart doing this? What do they stand to gain? It turns out that Wal-Mart could save itself billions of dollars, according to a report by A.T. Kearney. Savings will come from improved tracking of supplies, which should lead to a 5 percent reduction of store inventory requirements. Labor costs for inventory management are projected to drop by 7.5 percent. Wal-Mart expects initial labor efficiencies of 10%-20% at its distribution centers. A mere 1 percent reduction in out-of-stock would translate to $2.5 billion in added annual sales, since Wal-Mart's total annual sales is about $250 billion.

Wal-Mart handles approximately 8 billion cartons of goods annually, and its top 100 suppliers account for 12% of that. The RFID initiative therefore, will involve tagging and tracking millions of cartons beginning in 2005.

Wal-Mart RFID System Requirements

Requirements may be subject to change or revision in the coming months. Initial rollout includes these requirements:

Tags – Rather than wait, Wal-Mart has chosen to accept existing tags. They can be durable, temporary or permanent read-only

2004	2005	2006
Pharmaceutical tracking pilot	Top 100+ suppliers tagging for 3 Texas DC's	January – Next top 200 suppliers begin tagging cases and pallets
Refinements to RFID strategy		
May – 21 products from 8 suppliers tagged for Sanger, TX DC and 7 local Supercenters	June – 6 DC's, up to 250 Wal-Mart and SAM'S CLUBs	
	October – 13 DC's, up to 600 stores	

FIGURE 5.1

Implementation Timeline.

96-bit Class 0 (factory programmed), Class 0+ (WORM version) or Class 1 (WORM). Tags must be EPC compliant. Wal-Mart expects its suppliers to migrate to tags that support UHF Generation 2 (harmonized global standard) when performance, availability and price permit. Tags must operate in UHF (866-956 MHz) spectrum.

Tag application criteria – Tags are to be on both pallets and cases, including returnable containers, shrink wrapped bundles, bags and direct-to-store delivery trays. On full case pallet shipments, just the pallet tag will be read, not every case. A successful read of a pallet tag is defined as a minimum 3 reads at a distance of up to 10 feet from an antenna. Tag read rates of 100% are required. For individual cases, a 100% read rate is expected when cases are traveling up to 600 feet per minute, with a 6-inch separation between cases. Tags must be readable from any of 6 sides.

→ SEE PAGE 133
on reader antenna
placement.

Readers and antennas: Recommended are so-called agile readers that can handle multiple tag classes and frequencies. Readers must be Ethernet-based, have flexible output options, and RF environment awareness, built-in security, and be capable of disabling unused features (web servers, etc.). Dock portals require an antenna on each side of the dock door and an additional antenna above the door. Conveyor curtains require one antenna on each side of conveyor (speed up to 600 ft/sec) for case tagging. Cases must be read at 100 percent at 540 ft/sec.

Advanced Shipping Notice Integration

EPC data from case and pallet loads becomes part of the ASN a supplier sends to Wal-Mart ahead of the shipment. ASN involves instantaneous communication, via EDI, once packages and pallets are sealed and labeled by the manufacturer for shipping. The order detail contained in an advanced shipping notice associates case numbers with a specific pallet. It will be matched against the receiving detail (pallet ID code) that is automatically read at the dock door when a tagged shipment arrives (Fig. 5.2). A warehouse management system verifies receipt and directs the next step, such as checking the shipment into inventory or cross-docking for outbound transit.

⊙ SEE FIG. 5.2

OTHER RETAIL INDUSTRY RFID INITIATIVES

Although Wal-Mart assumed a leadership role by pushing its mandate first, other retailers are close behind and more are expected to make announcements.

Target Corporation, with approximately 1,200 stores in North America, expects top vendor partners to apply tags to all pallets and cases and start shipping to select regional distribution facilities beginning late spring 2005. Target's intent is to accept EPC tags from all vendors as a supplement to the current barcode markings at the carton and pallet level by spring 2007. Target sees RFID as a

FIGURE 5.2 Advanced Shipping Notice integration with RFID are part of the mandate.

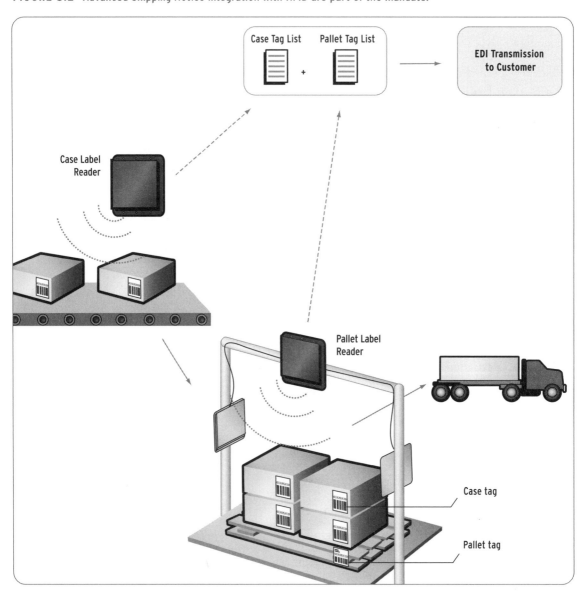

Case Tag List   Pallet Tag List

EDI Transmission to Customer

Case Label Reader

Pallet Label Reader

Case tag

Pallet tag

complement to current barcode and EDI technologies. Target supports a retail industry migration approach. For the foreseeable future, the current carton marking requirements for shipping containers will remain unchanged.

Albertsons is one of the world's largest food and drug retailers, with 19 DCs and approximately 2,300 retail stores in 31 states across the United States, including Albertsons, Jewel-Osco, Acme, Sav-on Drugs, Osco Drug, and Super Saver. The company is currently in the testing phase using RFID technology with select partners at the case and pallet level. Albertsons expects its top 100 suppliers to be participating in the RFID program at the case and pallet level by April 2005.

GLOBAL RETAIL INITIATIVES

Together with Intel Corporation, Europe's three largest retailers, TESCO PLC, Carrefour Group and Metro Group, formed the EPC Product Retail Users Group of Europe. This independent working group complements the efforts of EPCglobal, actively piloting EPC and RFID technologies in their supply chains.

TESCO, the largest retailer in the United Kingdom, and among the most active retailers testing RFID technology based on EPCglobal's work, will put RFID tags on cases and re-usable totes of nonfood items at its distribution centers and track them to stores. Goals are to rollout to over 900 stores during 2004. Tesco is using ETSI compliant 866 MHz tags.

Metro Group, based in Germany and Europe's third largest retailer, unveiled an RFID "Future Store," showcasing the benefits of the technology for shoppers. Metro Group has over 2000 stores in 28 countries. In their test store in Rheinberg, the company is implementing RFID within the store aisles and checkout areas as well as receiving. For inventory management, RFID is used on the shelves to detect low shelf stock, or misplaced items. Customers can scan products for themselves while in the store, and use an intelligent scale to calculate price on weighed items. Display terminals show customers information about products, including personalized ads and suggested recipes (Fig. 5.3). The checkout system quickly totals and displays the price of goods in their shopping card (Fig. 5.4). Customers can debit a credit card account or pay the cashier. A de-activator display after checkout allows customers to kill an item tag if they wish (Fig.5.5). Metro is making information on RFID, EPC and the Future Store concept available in the store to help educate their customers. Early surveys at the Future Store show that at least 25% of their customers have tested all the new systems themselves. Satisfaction with shopping experience has increased from 34% to 52%.

5.3A

5.3B

FIGURE 5.3A

Product information via RFID.

FIGURE 5.3B

Personal ads while shopping.

5.4

5.5

FIGURE 5.4

Checkout portal.

FIGURE 5.5

De-activator.

Photos courtesy of Metro

Carrefour, Europe's largest retailer and second largest in the world, sees global implementation of RFID as assurance that customers can have better product availability and value. Carrefour operates almost 10,400 stores in 30 countries across four formats: Hypermarket, Supermarket, Hard Discount and Convenience.

PHARMACEUTICAL INITIATIVES

Wal-Mart is currently piloting an RFID initiative in partnership with drug manufacturers, distributors, and under the watchful eye of the Food & Drug Administration. The goal of the pilot is to explore ways to eliminate drug counterfeiting. The initiative falls within a set of proposals and recommendations involving "track and trace" technology. RFID is viewed as having potential, because tags aren't easily tampered with, and the EPC code structure enables the assignment of unique identifiers throughout the process.

Prescription drugs have a rather complex supply chain, involving contract manufacturing, wholesaler intermediaries, and repackaging steps as drugs find their way to a retail counter. Counterfeiters have found multiple entry points in that chain to substitute drugs. The high retail value of certain breakthrough drugs has created high-risk categories, susceptible to counterfeiting. The counterfeiting problem is estimated at $7 to $26 billion within the $327 billion global drug market.

Electronic tracking technology, such as RFID, along with "chain of custody" business practices, could make it much more difficult for illegitimate and rogue operators to develop entry points within the distribution supply system. Participants in the Wal-Mart pilot have found that RFID has demonstrated cost effectiveness by improving inventory control, expediting delivery shipments and reducing product waste and diversion.

The recommended approach across the supply chain is called "one forward, one back." This approach is based on the concept that at each point in the supply chain, a certification is made that the drugs were received from a valid source (one back), and will be shipped to a valid source (one forward). RFID implementation is likely to be phased in, starting with the serialization of cases, pallets, and eventually smaller packages.

The Healthcare Distribution Management Association recently took the position that manufacturers and wholesalers should use EPC tags at the case level, with a goal for deployment of December 2005. In addition, HDMA is suggesting that drug packagers and manufacturers should set a goal of deploying EPC tags at the item level by 2007.

DEPARTMENT OF DEFENSE INITIATIVE

The Department of Defense has been using RFID over the past 10 years, mainly active tag systems for the identification of large containers. The DoD will begin requiring suppliers to use passive RFID tags on shipments as new contracts are issued to them starting in 2004, with the compliance of January 2005 set for initial suppliers. Full compliance is not expected to be achieved until 2010.

The aim of the initiative is to track cases and pallets of material, as well as high-value items tracked as assets with an assigned Unique Identification (UID) number. The DoD feels that the visibility gained will reduce safety stocks and improve forecasting. In addition, RFID deployment is seen as a way to re-deploy critical manpower resources to war-fighting functions, streamline business processes, and spread efficiency benefits to its supplier partners.

A final version policy is expected in mid-2004. The DoD is attempting to leverage EPCglobal standards, so as to share in the benefits of other initiatives including that of Wal-Mart. By making its mandate effective across the board, a great number of suppliers will be involved in 2005. See Figure 5.6.

Practically all of Wal-Mart's top 100 suppliers also supply to the DoD. There is a great deal of overlap with the rest. The DoD has approximately 43,000 suppliers, of which 500 are supplying most goods.

Of particular concern is the task of harmonizing code data, since it affects 1,500 logistics systems. In parallel to the Auto Id Center work, the DoD recognized the need for a unique identification (UID) number, and mandated the use of UID on all solicitations issued on or after January 1, 2004. The intent is for UID and EPC to be compatible.

The DoD already tracks cargo containers using active RFID tags. The chief of the DoD Logistics Automatic Identification Technologies office estimated that the military saved $300 million in Iraq through the improved visibility gained by RFID applications. The DoD is expected to stick with active RFID tags at the container and air pallet level, and to roll out passive tags for transport units and product packaging. The tagging requirements by RFID layer are shown in Table 5.1

⊙ **SEE TABLE. 5.1**

2004	2005	2010
Initial pilot projects – January	Suppliers begin UHF passive tagging of pallets and cases – January	43,000 suppliers achieve full compliance
Final policy – July		
Passive RFID requirement in all new contracts – August	Defense Logistics Agency distribution centers at San Joaquin, CA and Susquehanna, PA set up to use passive RFID.	

FIGURE 5.6

DoD implementation timeline.

RFID Tag Specification

The DoD will mandate the UHF generation 2 (harmonized) tag standard when it becomes available. Until that time, tests, pilots and initial implementations should proceed using the currently available EPC Class 0 and 1 tags. The overall goal is to embrace the open standard, EPC compliant tag along with the rest of industry, so as to leverage the economies and widely shared business practices that follow mass adoption.

Supplier Funding for Implementation

The DoD recognizes the cost burden placed on its suppliers, and expects to approve contract changes to absorb some costs. According to Attachment 2 of the Policy letter dated February 20, 2004, "Working Capital Fund activities providing this support will

TABLE 5.1

Tagging requirements by RFID Layer.

RFID Layer	Description	Tag Type	Class Tag	Frequency	Read Range	Starting January 1, 2005
0	Item	Passive	0, 1 or higher	UHF	3 m required	Not yet
1	Item package	Passive	0, 1 or higher	UHF	3 m	Required on UID & specified items
2	Transport Unit, Case	Passive	0, 1 or higher	UHF	3 m	Required
3	Unit Load, pallet	Passive	0, 1 or higher	UHF	3 m	Required

use the most current DoD guidance in determining whether operating cost authority (OA) or capital investment program (CIP)
authority will be used to procure the required RFID equipment."

Sources and Further Information

Wal-Mart's Roadmap to RFID, by Dave Kelly, *RIS News,* January 2004.

Meeting the Retail RFID Mandate, white paper, AT Kearney,
November 2003.

Memorandum, Radio Frequency Identification (RFID) Policy,
Michael Wynne, Acting Deputy Under Secretary of Defense,
Acquisition, Technology and Logistics, DoD, October 2, 2003.

*Memorandum, Radio Frequency Identification (RFID) Policy –
UPDATE,* Michael Wynne, Acting Deputy Under Secretary of
Defense, Acquisition, Technology and Logistics, DoD, February 20,
2004.

*Memorandum, Policy for Unique Identification (UID) of Tangible
Items – New Equipment, Major Modifications, and Reprocurements
of Equipment and Spares –* USD (AT&L), July 29, 2003.

Understanding the Wal-Mart Initiative, white paper, Matrics
Systems Corporation, 2004.

Metro Store Future, presentation by Gerd Wolfram, Director of IT Strategy, IT Buying and Development Services, Metro MGI Information Technology, RFID Journal Live Conference, April 2004.

CHAPTER 6

Smart Start to RFID - 4 Friendly Phases

IN THIS CHAPTER:
Initiating a smart label pilot.

Even if it weren't a requirement for doing business with Wal-Mart and DoD in 2005, RFID is poised to transform supply chain operations over the next decade. Using smart labels allows every element along the supply chain – from the manufacturing floor, to warehouses, to store shelves – to track what the product is, when it was made, and where it is going. The information contained in the smart label will improve efficiencies by reducing errors in receiving, keeping product in stock, decreasing misplaced inventory, theft and counterfeiting, and lowering administrative and labor costs. Ultimately, RFID can ensure products are on the shelves when customers want them, increasing revenue for manufacturers and retailers alike.

PHASE 1: GETTING STARTED

Along with the promise of RFID come the challenges in implementing it. Navigating your way through a complicated new system that requires its own hardware and software is a daunting mission. Combined with the complexities of evolving standards, converting today's barcodes to tomorrow's electronic product codes (EPC), and the prospect of how all of this changes the way your company functions – it is easy to understand why you might take a long look before you make the leap to RFID. But there are benefits for those who embrace the technology now, and it is possible to start smoothly and slowly, one step at a time through an achievable four-phase plan.

Set Up a Development Environment

You will want to set up a development environment for small-scale, controlled testing. As you assemble your test lab, it's valuable to make decisions with the future in mind. If you think of your integration and vendor partners as the long-term RFID team that will successfully guide you through deployment, you can plan accordingly. To help make the right choices, the following can aid you in your partner selection:

→ SEE PAGE 129 on lab testing.

Choose Smart Partners

In this early adopter, emerging technology phase of RFID, there are many vendors jumping into the marketplace with promises of expertise to help with your RFID deployment. Therefore, it's more important than ever to focus on technology-based and solutions-based companies that can help you migrate and upgrade in a timely way through each level in your implementation process. It's essential that external partners on your pilot team are leaders in the field and participating members of EPCglobal, the not-for-profit standards organization leading the adoption and implementation of the EPC network. EPCglobal members work hand-in-hand with the leading retailers, suppliers and DoD. As a participant in EPCglobal, your RFID partner can better understand your business requirements for effective implementation, can influence developments and progress by integrating technical, cost and performance needs into standards, and will keep your organization's program up-to-date. Look for vendors who can provide a fully integrated, cost-effective development environment along with:

Experience and core competencies – Your software and hardware partners should be technology-driven and market-driven companies, as opposed to product-driven, with a successful track record of offering auto-ID solutions specifically designed for the supply chain, such as compliance. Your ideal partner will have in-depth experience that goes beyond simply a product offering and beyond RFID. Also note if they are early adopters of the RFID technology, have experience specifically related to the Wal-Mart and DoD mandate, and are directly involved in numerous pilot programs.

Solution builder commitment – Success starts at the top. As you analyze RFID partners, ask what their management's vision is of RFID solutions. Is RFID a company priority, supported by the entire management team and funded for development? Do they provide an end-to-end solution? Does their organization have a professional services group that can help with the seamless integration into your enterprise network without business interruption?

After sales support – Support and maintenance issues continue long after the initial installation. Your partners' product and application engineers should engage in an ongoing dialogue, answering your questions, assisting you with ways to accomplish your RFID objectives, and listening to your feedback. They should continue to provide technical support throughout your program, helping with the full range of technical concerns, from integration, to spare parts, to trouble-shooting the entire system.

Piloting Smart Labels

Since labels carry the passwords that get your carton into the warehouse, they are the logical starting point for your pilot program. With a Printronix Smart Label Developer's Kit, you can produce smart labels right out of the box. The kit includes software migration tools that provide a seamless transition to encoding and printing smart labels without incurring high reprogramming costs, waiting for official EPC numbers, or changing anything within your front end or back end applications.

→ SEE PAGE 176
for smart label developers kit.

The kit includes a suite of applications that convert standard UPC and Global Trade Item Number data from barcode print data and allow you to simultaneously print and encode them into the RFID tag. The applications' flexibility allows you to select from many common shipping label templates such as ITF-14 and UCC/EAN-128. And while the codes printed on the labels won't be the final EPC (as standards are still being developed), they can be written, printed and verified for this initial testing phase.

Label Testing – Now that you can encode smart labels without having to wait for EPC numbers, you have the means to test read ranges, read speeds and data capture. You can determine the distance from which the labels can be read, whether the products themselves affect RF signals, where you should locate the label on the carton, and variations to read angle and distance. As you become familiar with optimum read speeds and work out the intricacies of capturing and reading data, you will, most importantly, arrive at solutions to improving and maximizing system accuracy and efficiency.

Label Placement

⊙ SEE PAGE 127
on case analysis.

Package contents and label configuration, design, space and angle all can make a difference between a 100% read rate and a 0% read rate (Fig. 6.1). You will need to keep these factors in mind as you determine the placement of your smart label on the case or pallet.

Package design – Adding a smart label to cartons and packages that were designed by marketing to achieve brand objectives may limit available physical label space. Determining placement is an exercise in problem solving and trial and error as you contend with existing package and label sizes and graphics.

Label requirements – In large products such as a television or printer, the carton it comes in will be tagged. The smart label will identify the individual item, advancing the identification to item level marking.

Package contents – Liquids and metals absorb or reflect radio waves. Careful application of smart labels is required for items such as foil-

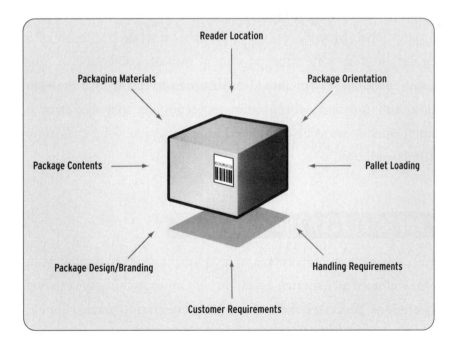

FIGURE 6.1

Label placement depends on a number of factors.

bagged chips, liquid detergent, and canned goods. Often the effective area is extremely limited and it will take extensive experimentation.

Good and Quiet Labels

A label is considered good when the RFID data is written to the tag correctly, the correct image is printed, and content data is verified against the source. If the printed and encoded data can't be verified against the source, the label is considered defective and voided from the system (Fig. 6.2). To ensure that no EPC numbers are lost, the printer should be programmed to clearly overstrike and void the defective label and print another label using the same EPC data. When a verified tag can't be read from a normal distance, it's called a quiet label. In some cases a quiet label may be the result of a defect in a specific label within a roll of good labels. Your print/encoding system should be designed to distinguish between quiet and non-quiet labels, removing yet another source of error. A quiet label needs to be eliminated from use if you want to achieve 100% read rates.

PHASE 2: TEST AND VALIDATION

→ SEE PAGE 129
on location testing.

RFID is uncharted territory, and it's a long journey to deployment. You will need an experienced guide who understands your existing operations, processes and systems. Your integration partner should not only have RFID expertise, but should also have the industry knowledge to help you develop an implementation plan that defines all workflow tasks, responsibilities, milestones and related costs, and assist in establishing realistic performance targets. The integrator is your co-pilot, so test their background in supply chain solutions, look at their credentials in technology, and ask about their work relating to the Wal-Mart and DoD mandate.

A label is considered good when the RFID data is written to the tag correctly, the correct image is printed, and content data is verified agaiinst the source.

If the printed and encoded data can't be verified against the source, the label is considered defective and voided from the system.

FIGURE 6.2

Examples of good and bad (quiet) labels.

→ SEE PAGE 161

on smart labeling systems

integration.

System Integration

In this phase of test and validation, it is important to begin learning how RFID devices will integrate with your (Enterprise Resource Planning) and WMS (Warehouse Management Systems). Testing and evaluation allow you to preview the extent of capabilities RFID brings to your enterprise and the supply chain. Because RFID supports such areas of your business as resource planning, parts purchasing, order tracking, customer service, inventory management, transportation management and accounting – by providing extensive, real-time, accurate information – it's predicted that you will realize significant gains in efficiency. Early in your vendor partner selection, note that for smooth sailing in this phase, select equipment that fully integrates with leading WMS and ERP suppliers. For software, look for complete turnkey solutions such as Manhattan Associates' RFID-in-a-Box.

Simulate a dock door and a conveyor using fixed-mount readers as the Smart Label Developer's Kit creates a sampling of labels for typical cartons and pallets. (You might also choose to begin your testing in the established lab of an integrator partner for this initial phase.) Products such as Alien Technology's RFID Development Kit with reader, antenna and development system software will accelerate your progress through this phase.

Validate Vendor Choices

As you approach your pilot program implementation, evaluate how your equipment is working for you. No matter what manufacturer you team with, there are expectations you should maintain for RFID. For example, make sure your printer partner offers:

- Complete encoding solutions
- RFID extensions and drivers
- Ability to extend your development environment to more than one printer to support pilot runs of 10,000 to 50,000 smart labels.
- Certified smart labels in unlimited quantities
- Rapid development capability for your unique label design

Printers and labels need to work together with the RFID equipment, communicating information back into the ERP or WMS system. Reader partners should be able to provide RFID solutions for various global frequency requirements as they evolve. During the implementation phase, readers can be positioned depending on your needs at various locations such as shipping and receiving dock doors, product routing conveyors, picking and sorting configurations, and forklifts. Handheld readers facilitate inventory counting, locating and reconciliation, and should have the ability to capture both barcode and RFID.

For your reader partner, look for vendors who can offer solutions to your various geographical and form factor requirements.

Support services are always available to assist with product documentation and integration updates, repairs, spare parts and technical support requirements.

For all vendors, make it a point to find out their support strategy, and ask if they have established alliances with market-leading partners in RFID project planning and deployment strategies.

Confirm Label Type and Placement

During test and validation, solve where to position labels on different product types and how to apply as volume increases. As an example, in the case of one particular bottled drink, tag placement for 100% read rate was critical to within 1/4".

PHASE 3: PILOT IMPLEMENTATION

The objective of the pilot program is to develop a predictable and scalable system. This requires you to achieve precision in placement, output and performance. Careful measurement and documentation throughout this phase will facilitate problem solving with your partners and selected customers, to ultimately eliminate errors and establish processes. You will want to mark critical milestones to chart the development of your system. Along the way, stop and assess your solutions – is smart label placement formulated and confirmed for different products? Should you run parallel pilots for different divisions of your business because of significant differences in processes? Is it time to incorporate additional smart label printers to your system? The pilot phase is the time to tool up for handling greater volumes with real-life criteria in actual working environments. You will build knowledge and confidence in the system as you work out the everyday demands faced by your business, even though you are applying the tests only to a limited volume. Figure 6.3 lists the steps for this and other phases.

⊘ SEE FIG. 6.3

In order to achieve the pilot objectives, you will want to:

- Set up equipment in other facilities/divisions to discover and solve any anomalies within each facility.
- Verify your ability to capture and transfer data and send it between locations.
- Capture data on one specific product out of a test run of assorted SKUs.
- Educate employees on the importance of the RFID system and how it will affect the way they work. If tags are being applied manually, this is a critical part of the learning curve.
- Partner with a retailer to send test shipments to verify system compatibility.
- Subject the system to the rigors of a typical production or shipping facility.
- Handle higher volumes (50,000 or more).

• Measure results to test the viability of larger scale.

• Work with your partner team to eliminate errors.

• Consider expanding your pilots to additional products or geographies after successful completion of your first. You may find that different divisions or product lines require different pilots.

You will have successfully accomplished your pilot implementation if you have:

• Measurement of results, including establishment of performance metrics.

• Integration of ERP/WMS to extract the data out of the label and

FIGURE 6.3

Steps for implementation.

RFID IMPLEMENTATION PROCESS			
❶ GETTING STARTED	**❷ TEST AND VALIDATE**	**❸ PILOT**	**❹ IMPLEMENTATION**
Assemble your lab	Involve a knowledgeable systems integrator	Develop a predictable and scalable system	Explore opportunities for new efficiencies
Set up a development environment	Evaluate/test various software applications	Set up equipment in other facilities divisions	Capture and manage data
Focus on technology and solution-based smart labels	Evaluate/test with warehouse infrastructure	Verify your ability to capture and transfer data between locations	Implement RFID network and device management
Start making smart labels	Test read ranges, read speeds and data capture	Capture data by specific SKUs on a run of assorted SKUs	Deploy smart media management
		Measure results	

pass the information back to the system for operations management.

- Defined different label and antenna requirements for different SKUs.

- Programmed your system to detect human errors in tasks such as label selection and placement as your equipment checks the data and media and alerts you to problems. Your systems and processes will become systems and processes error-proof.

- Decided: manual or automated application of smart labels, in-process or post-process?

On January 1, 2005, only selected suppliers and only those products shipping to three Wal-Mart distribution centers in Texas are required to have full RFID compliance. It may be more cost effective for the year 2005 to limit your RFID compliance to only those products shipping to those specific warehouses. If you are running out of time, consider the feasibility of making the application of smart labels a post-production step rather than an integral part of the manufacturing process. De-palletizing cases, adding smart labels and re-palletizing might be a viable short-term solution. The safest and best way to help you make this decision is to consider three factors: 1) volume, 2) number and percentage SKUs you need RFID on, and 3) how extensive your current automated process is.

⊙ SEE PAGE 147 for slap & ship with EPC management.

Solving RFID implementation issues, even if the requirement is for a small percentage of your shipments the first year, will provide a strong foundation for the following year when 100% of Wal-Mart shipments require smart labels, and DoD requirements become effective. By the end of Phase 3, you will have locked down your business processes and procedures, tested software and hardware, and verified your system accuracy at higher volumes and speeds. For scheduling purposes, Phase 3 should be completed by the end of Q3, 2004.

PHASE 4: IMPLEMENTATION

Although full RFID deployment is still ahead of us, hundreds of companies are on their way. Whether you are just starting or already piloting, several issues considered in your early decision-making will facilitate a more efficient, successful implementation.

Ensure Against Obsolescence

Technology is evolving, standards are not resolved, and protocols will change. Smart labels will migrate to UHF Gen 2. These changes in the industry will mean changes in your equipment. So choose a vendor offering asset protection to protect your investment, with upgradeable firmware (for example, to support new standards or data formats) and scalable solutions so you won't have to start over. We suggest that you ask a set of pertinent questions to help you make informed decisions about product and vendor selection:

• How many pilots are they working on?

• Can they articulate examples and case histories from their experiences related to the Wal-Mart requirement?

• Do they have a strong list of RFID partners?

• Are they a global company to manage your international locations?

• Do they offer formally organized professional services such as label design and verification, on-site assessments, training, and integration and migration consulting?

ROI Lies Beyond Mere Compliance

During the implementation phase, you will explore opportunities for new efficiencies and build metrics into your processes to quantify improvements, forming a foundation for ROI. This reinforces that the solutions you pick for pilot runs need to be scalable, robust and industrial strength for cost-effective deployment. And even if your processes include manual application of labels for shipping (slap and ship) at this point, it's important to keep your future automation capabilities in mind as your system expands. This is a critical factor when choosing your printing solution.

Deploying smart printers will provide these building blocks to ROI:

Validation and verification – With validation built into printing equipment, you can correlate 100% barcode reads back to 100%

RFID reads and be able to cross reference. Without manual intervention, your system will be able to check every label against your database to verify that what you read on the label is actually what it should read. If there is a discrepancy, it will immediately back up, cancel and overstrike the label, and print a correct replacement. Read-after-print quality control is designed to eliminate defective labels from entering the supply chain. It will also prevent print production slow-down, minimize the cost of labor, and avert product returns and fines for non-scannable labels.

Data capture – The ability to archive information for enterprise management brings the highest level of visibility to your operations. Through data capture, nearly instantaneous visibility of supply chain activity allows you to make more accurate sales projections and purchasing decisions. EPC data, once integrated into your database, can provide time, location, and batch information that when passed back into the system can identify and locate specific products at any point in the supply chain.

→ SEE PAGE 166 on enterprise-wide smart label print management.

Network and device management – Access to real-time information and control of your devices improve efficiency and productivity and help you make informed decisions in managing them. Network print management systems will provide instantaneous visibility to every discovered device and allow users to simultaneously configure an unlimited number of printers. These print management solutions will also support management of the additional RFID encoder capabilities. Providing instant visibility (enterprise view), instant notification through e-mail alerts and pages, and remote diagnosis, these tools will enable you to send test results over the printer network for viewing and storing in an XML file (or other formats) for later comparison with the data stream sent to the printer.

Smart media management – Your application will police itself and alert you if something is wrong. If the EPC doesn't correlate product to label, you will be notified immediately. Proactive detection will ensure the label placement is right, the class of label is right, and the antenna design is right for the label.

Industrial design – Your equipment is an investment in the future. To support you from test to pilot to deployment, your printer will need to be rugged and dependable, able to handle growing volume. Remember that the cost of a printer itself in the overall scheme of things is insignificant relative to the investment in infrastructure, tags and costs of non-conformance and noncompliance as you consider the downtime, lost productivity, fines and product returns.

Smart Labels Will Get Smarter

Developments in RFID technology will continue to yield larger memory capacities, wider reading ranges, and faster processing.

Although the cost of RF chips prevents RFID from replacing barcodes any time soon, the technology will continue to flourish because of its interactive capabilities. Growth and speed will increase exponentially, creating capabilities that will have a major impact on supply chain operations over the next few years. Adopting RFID today affords you the opportunity to grow with the technology, to define and refine your system and realize the benefits sooner.

RFID PARTNER CHECKLIST

Consider the following requirements for your integration and vendor partners:

____ Participating member of EPCglobal

____ Partnered with other experienced RFID partners

____ Can provide upgrade paths

____ Offers asset protection program

____ Upgradeable firmware

____ Scalable solutions

____ Robust industrial equipment

____ Experienced in RFID specifically related to the Wal-Mart mandate

____ Involved in other pilots, can clearly articulate case study-type examples

___ Able to validate and verify without user intervention

___ Offer an enterprise network management solution

___ Formal professional services organization

___ Label verification testing, organize data and files for clients

___ Training, integration and implementation consultants

___ Global organizations to support clients' international locations and expansions

___ Able to supply volume quantities when needed

___ Equipment fully integrates with major supply chain software and other enterprise programs

Sources and Further Reference

List of Printronix RFID Partners, available: www.printronix.com.

CHAPTER 7

Case Analysis

Case analysis is the process of evaluating packaging in order to determine where to place an RFID tag or smart label. The objective of case analysis is to achieve optimum tag readability in all the circumstances where the tag will be read. Case analysis should occur before you decide how the tag will be applied. Since you may be applying smart labels to cases by a variety of methods, its best to determine what and where first.

A case analysis should answer these questions for you:

1. What are the compliance requirements and implementation schedule?
2. What kinds of products and packages require labeling?
3. What are the packaging volumes?
4. What types of smart labels are needed?
5. What are the package label placement locations and tolerances for consistent RFID read rates?
6. What is the effective read range?
7. What is the palletizing strategy?
8. Will RFID work within my existing product packaging, palletizing and shipping process?
9. What training will be required for packaging, warehousing and distribution center workers?
10. What costs are involved with each label application method?

LAB TESTING

The best place to start case analysis is off line, in a non-production environment such as a lab. In your lab you can start with a known good tag and reader, and determine how packages and package contents will affect RFID tag encoding and reading. You might want to consider the services of a systems integration company that offers RFID case analysis. They may have a facility already set up to conduct the lab testing phase.

Lab testing should be used to evaluate and test the following:
- Product chemistry, packaging types and composition, and how they affect RFID
- Tag and label choices, costs and availability
- Label placement options and tolerances
- Tag encoding methods
- Reader and antenna combinations
- Read rates at various line speeds and distances.
- Pallet loading

LOCATION TESTING

Evaluation and testing continues as you move into the pilot and production phases. Questions that are answered during these phases include:

Divert or integrate? – Whether you create a separate compliance process and divert production to it, or integrate RFID into your standard packaging line, involves a number of factors. They include existing production and packaging methods, the types and volumes of product that falls under a compliance mandate, read rates and investment timing.

RFID portals – The selection and setup of packaging line and dock portals involves an analysis of the physical environment within the reading area, the evaluation of various reader and antenna configurations, and performance testing.

Short, medium and long term approaches – You should be able to determine the capability and capacity constraints of your initial approach, and begin planning for the medium and long term.

LABEL PLACEMENT

The location and orientation of a smart label on a case or pallet can be critical. Product composition, package geometry, packaging materials, pallet loading, proximity and orientation with respect to the reader antenna are all variables that have to be considered. In pilot applications for smart labels on packages containing liquids, label placement variation of as little as 0.25" (6 mm) from optimum has been found to affect read rates.

Tag presentation – As a tag passes through the read window, ideally it should be on the same plane as the antenna. The flat face of the tag should be parallel to the flat face of the antenna (Fig. 7.1). If a linear polarization antenna is being used, the tag must be oriented vertically or horizontally with respect to the polarization. See Figure 7.2.

→ SEE FIG. 7.2

Tag coupling – Tags placed on or very near metal objects, such as aluminum cans or foil, may electrically couple with them. This may short out the antenna. Proper tag selection and placement is very important with packages containing metals.

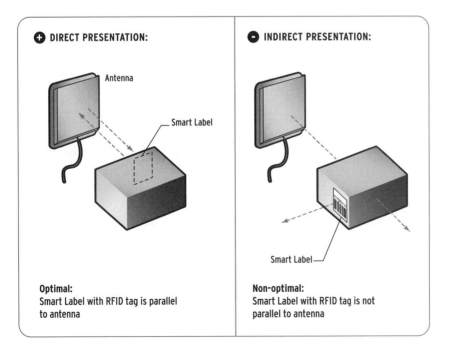

DIRECT PRESENTATION:

Antenna

Smart Label

Optimal:
Smart Label with RFID tag is parallel to antenna

INDIRECT PRESENTATION:

Smart Label

Non-optimal:
Smart Label with RFID tag is not parallel to antenna

FIGURE 7.1

Ideal tag presentation is where the flat face of the tag is parallel and on the same plane as the flat face of the antenna.

→ SEE PAGE 91
Wal-Mart requirements.

Pallet testing – You may find that interior cases on a pallet will not read and a new pallet loading strategy is necessary. If 50 cases are on a pallet and only 40 tags register themselves with the reader, you will need to find the quiet tags. An obvious aid to troubleshooting is having a the tag code printed on the smart label, in bar code and human readable form. By reading the bar code and correlating label data to the tags, the individual cases with quiet tags can be identified.

FIGURE 7.2

Label placement should align with antenna polarization.

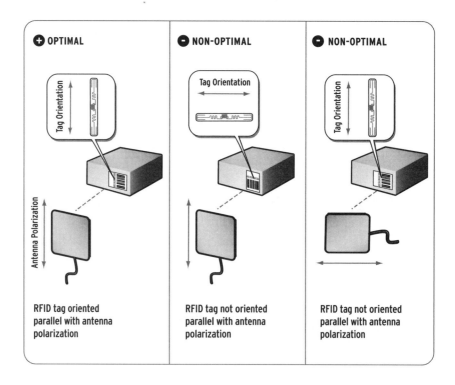

READER ANTENNA PLACEMENT

Since the power output of a reader is regulated and fixed, antenna design and placement is perhaps the most important way to tune the RF signal to an environment. Varying the reader antenna placement is usually the easiest troubleshooting step, but may be the trickiest one to do well. Figure 7.3 shows a typical dock door layout.

There are over a hundred different antenna designs for HF and UHF systems. Gain, efficiency and radiation pattern are variables

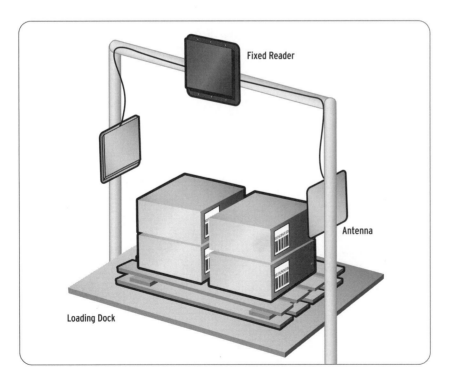

FIGURE 7.3
Dock door antenna placement.

that can be addressed by the design of a reading zone, and the selection of an antenna or antenna array. In general, antennas emit two types of energy patterns, linear polarization (Fig. 7.4) and circular polarization.

A useful tool for antenna selection and placement is a mapping of its radiation pattern within the operating environment. See Figure 7.5. This can be done by placing a known good tag at various points in the read area, and attempting to read the tag. By marking out a grid pattern on a floor diagram, indicating where the tag does and does not read, you can determine the practical range of the antenna. Since read redundancy (3 or more good reads within a short

FIGURE 7.4

Linear and circular polarization.

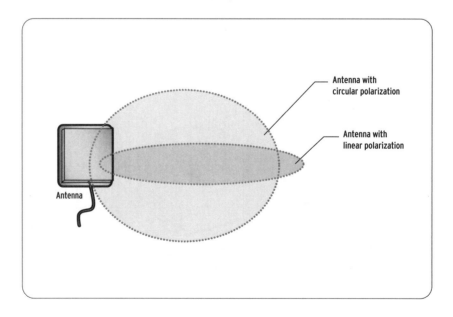

Antenna with circular polarization

Antenna with linear polarization

Antenna

timeframe) is an important performance characteristic, especially if the reader or tag will be moving relative to each other, you should determine the limits of the read area. Testing software and tools are available from tag manufacturers to help map read areas and determine read redundancy.

CHARACTERISTICS OF RF AFFECTING READ RATE

Passive-tag RFID is a new technology compared to bar codes. There are relatively few retail supply chain implementations, and few

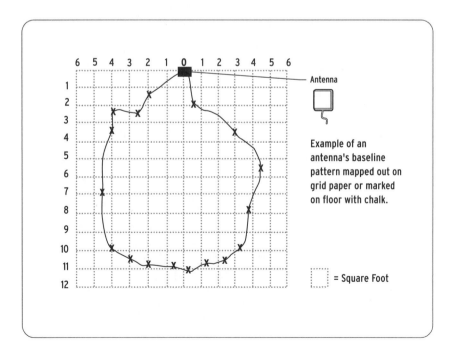

FIGURE 7.5

A floor diagram with tag reading areas boundaries marked.

Antenna

Example of an antenna's baseline pattern mapped out on grid paper or marked on floor with chalk.

□ = Square Foot

packaging engineers and system integrators with actual application experience. In contrast, many years have gone into the development of bar code printing techniques and the engineering of high speed packaging lines with bar code verification.

The hard science of RFID is just beginning to come to terms with the environment of a retail distribution center. It is likely that every DC will be different enough to make the troubleshooting of read rate issues feel more like an art than a science, at least at first. The following are characteristics unique to RF that affect read rates:

Translucence – Some materials offer little or no barrier to RF energy passing through them. Clothing made of organic and synthetic fibers, paper products, wood, non-conductive plastic and cardboard are translucent to RF. Paper packaging with foil lining, however, may block RF.

Absorption – Liquids, materials containing liquids such as foods, and liquids and foods containing salts in particular, will absorb UHF. Carbon containing compounds, such as graphite in solid or powder form, will also absorb UHF. What absorption does is attenuate, or weaken, the electromagnetic field propagating from a reader antenna or back from a tag antenna. Figure 7.6 is an example. Absorption varies by substance and by the frequency of a signal. It is possible to calculate the absorption rate of various substances to a certain frequency, and the resulting dielectric loss.

Shielding – Metals and very thin metals foils particularly can conduct a radio wave away from a target, not allowing it to pass through. Shielding material can behave like an induction coil, moving electrons in parallel with the induced current in a tag antenna, creating an opposing field that weakens the signal. See Figure 7.7. In general, higher radio frequencies are more easily shielded than lower frequencies.

Detuning – Tag antennas, are greatly affected by their immediate surroundings. A tag attached to a case of soda, for instance, is going to be more affected by its location (top of case, bottom of case, etc.) than anything else. Absorption and shielding from the cans will reduce the amount of energy that reaches the tag, and reduce the backscatter signal going back to the reader. Tags that are placed too

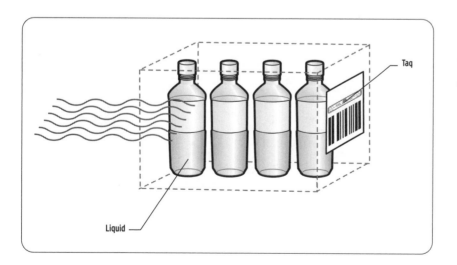

FIGURE 7.6

Liquids tend to absorb and weaken radio waves.

close together can capacitively couple to one another, detuning their antennas. The metal on conveyors, forklifts and other handling equipment can also detune signals, by blocking and reflecting them. Tags with suitable antenna geometries, proper placement on individual cases, and proper case orientation on a pallet can improve read rates. Package re-engineering may also be required.

Reflection – At UHF frequencies, signal reflection is possibly the most important fundamental problem for RFID. Because of reflections, a reader signal may not penetrate a shrink-wrapped pallet, for example, and the tags never receive enough energy to turn on. Reflections are due to the surface of a material having a different dielectric constant from that of the surrounding air (Fig. 7.8).

FIGURE 7.7

Coiled wire could shield an RF tag by conducting energy away from it.

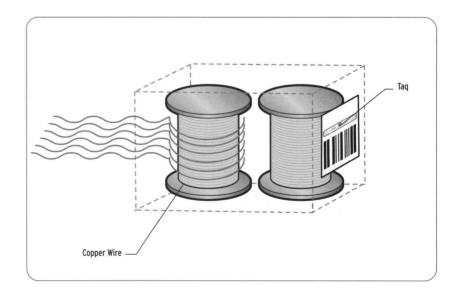

Tag

Copper Wire

Interference – Interference creates so-called "dead zones" due to the geometry of the environment. Conveyor apparatus can induce dead zones through vibration or EM discharge from motors or controllers. Other RFID systems, wireless computers, radios, and phones all can create interference, but it is usually filtered out through the reader/tag air protocol. Electrostatic discharge, from materials that accumulate static electricity and are not properly grounded, can also create interference. A reader signal can interfere with itself because of multiple reflections from other material surfaces. Examples include reflections from surfaces as a signal goes through a narrow opening to reach a tag (Figure 7.9), or a signal that bounces off of a metal object and reaches a tag nearly simultaneously (Figure 7.10).

 SEE FIG. 7.9

 SEE FIG. 7.10

FIGURE 7.8
Reflection of radio waves.

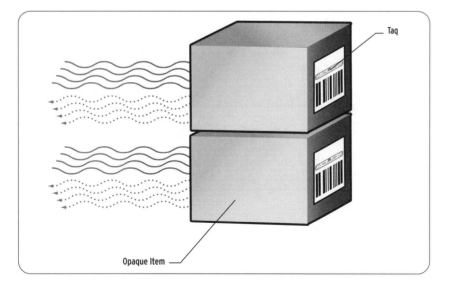

Taq

Opaque Item

FIGURE 7.9

Clipped signal due to interfering surfaces.

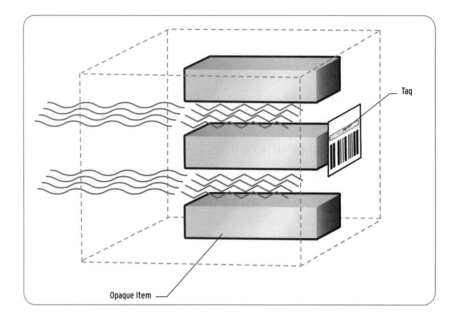

Taq

Opaque Item

FIGURE 7.10

Reader signal interfering with itself.

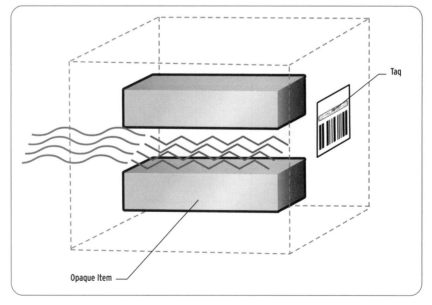

Taq

Opaque Item

Achieving acceptable performance for a retail supply chain RFID application may require a multi-disciplined approach. Process performance analysis tools such as affinity diagrams (See Figure 7.11 for an example), Pareto analysis and design of experiments, may help identify the sources of problems. In may be necessary to repeat the case analysis activities with the help of a system integration partner.

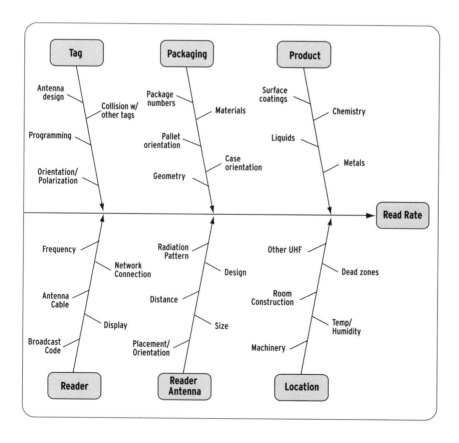

FIGURE 7.11

Example affinity diagram for troubleshooting read rate.

Sources and Further Reference

RFID Primer, white paper, Alien Technology Corporation, 2002.

Packaging and RFID Special Interest Group, available:
www.autoidlabs.mit.edu.

CHAPTER 8

Smart Labeling Approaches

Co-Authored by Rick Fox, Fox IV Technologies, Inc.

IN THIS CHAPTER:

How to apply smart labels to meet compliance and production requirements.

Approaches to smart labeling are based on when a label needs to be applied, where on the case, and at what rate. Before any labeling approaches are to be considered, a case analysis should be conducted to resolve RFID tag issues. If your case analysis activity is successful, you have identified where to apply labels, and the best tag antenna and reader antenna configurations to use. Now you can look at employing one or more application approaches to meet either production or compliance requirements.

A fundamental principle of package engineering is to conform product to supply chain capabilities, not the other way around. Products come in all weights, shapes, sizes, composition and condition; packaging engineers put them in conforming shapes so that they can be handled with little exception by standard material handling machinery and transport methods. This streamlines the movement of product through the supply chain and guarantees that it reaches the customer in ideal condition. Smart labeling follows this same principle by conforming RFID to supply chain requirements.

The right approach depends on many factors. What may seem to be the best approach initially might upon further investigation turn out to be a wrong turn and a costly detour. Let's categorize and briefly review the approaches:

Tags integrated with packaging – Disposable corrugated packaging with built-in RFID has several attractive characteristics, most notably that it eliminates tag handling during packing and sealing. Encoding can occur before or after packing. Tag acquisition and integration costs are pushed down to your suppliers. Disadvantages are many, however. Industry analysts predict that it will take 3 or 4 years for packaging companies to overcome the physical hurdles and make available a comprehensive offering. Error recovery, rework, and charge-backs may actually drive up costs. If you have multiple product lines that require different tag types and placement, you will have to purchase, inventory and manage a greater selection of package types, and match them with product. Lastly, paperboard packaging with embedded tags may have environmental impact and recycling issues that may be subject to green laws that may vary state to state and country to country.

Apply, then encode tags – Adhesive-backed RFID tags are an alternative in instances where cases are pre-printed with a bar code. The apply then encode process involves using an applicator upstream to affix the tag to a case, and an overhead reader to encode the tag as it moves through the packaging line. This approach could streamline the process and drive down costs, especially if existing applicator equipment can handle the tagging task. A programmable applicator could vary tag placement based on package contents. One drawback is that there is no way to validate or detect a quiet or bad tag prior to applying it to a case. This could cause a significant build-up of cases that require rework. Most label applicators in use today have not been designed to accept rolls of tags. Tight turns and pinch points in the tag carrier path could cause tags to lift and jam the applicator, and severe damage to the tags. In addition, poor adhesion could cause tags to fall off with no clear visible indication of the tag being missing. The labeling industry has spent years perfecting adhesives for various applications; knowledge that many tag-only suppliers are just starting to acquire. As a result, this approach may add variability to the process, by increasing the potential for quiet or bad tags, unprotected tags getting damaged or shedding from cases, and encoding problems.

Encode, then apply tags – A way to address some of the shortcomings of apply then encode is to encode the tags first, using an encoder built into the applicator. Unfortunately, approaches to this process are relatively unproven as of the date of this publication. On-pitch tag stock offers less flexibility and performance than smart label stock. In addition the tags by themselves are exposed and at risk of shedding because of limited choices for adhesives. Lastly, this approach does not include the clear marking of a case with an EPCglobal logo to meet guidelines on RFID identification for consumers. Consumer notices pre-printed on containers could be subject to change as marking standards continue to evolve.

Encode, print, then apply – Smart labels follow this approach, where an applicator is directed to encode and print a label on demand and just before applying it to a case. Validation, error-detection and recovery capabilities ensure that the tag is good, it matches the human readable information, and application can occur at the appropriate point in the packaging process. This approach can also be used for encode, then apply applications, by not printing the label and using the label as a carrier for the tag.

"SLAP & SHIP"

Slap & ship is distribution center language for separating a label from its adhesive backing and affixing it by hand to a case or pallet. This approach can be labor intensive and slow. It runs counter to the long-term goals of the speed and labor savings vision of RFID technology. Slap & ship is where some RFID applications start, however, in the lab and in early pilots, to verify placement, read rates, read speeds and other issues. A compliance labeling application can be initially set up as slap & ship with EPC management, as described in the next section, then converted to print & apply with the addition of an integrated RFID printer/applicator.

Figure 8.1 shows all of the locations in a manufacturing or distribution center where a slap & ship approach may have value. It is appropriate for reject lines, returns, special handling, or where the RFID

⊙ SEE FIG. 8.1

FIGURE 8.1

Locations for slap & ship labeling for RFID.

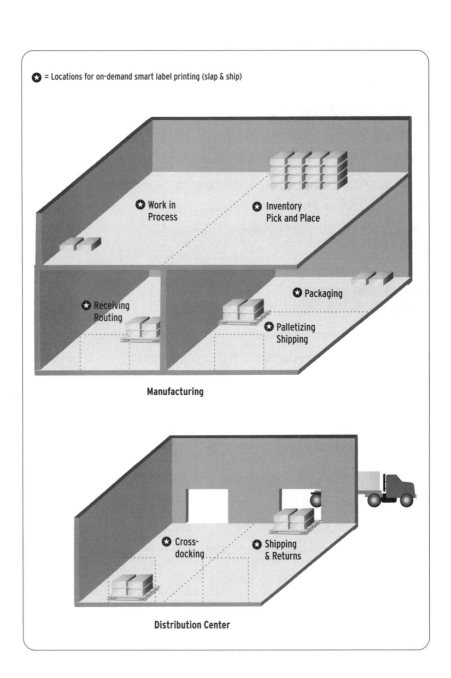

requirement does not justify a full production line implementation.

Table 8.1 compares some of the pros and cons of slap & ship. Most distribution centers will need a slap & ship capability for on-demand encoding and printing of labels.

SLAP & SHIP WITH EPC MANAGEMENT

A slap & ship approach can help a company get started with compliance labeling in a productive way. Products to be shipped with RFID labels would be separated and diverted to a staging area, particularly if production line speeds prohibit slap & ship. The diversion can be achieved by splitting a conveyor, or by delivering

Advantages	Disadvantages
Most flexible approach	Labor intensive
Least engineering content & cost	Not integrated with process
Allows tagging at distribution center just before shipment	Does not provide upstream benefits of RFID
Fast start-up	Low volume
Back-up for methods that are brought on-line	Does not scale
Achieves initial compliance labeling requirements	Lengthens investment period
Potentially least disruptive approach	May require separate staging area
On-demand	Potential for operator error
Allows company to learn about RFID	Long-term competitive disadvantage

TABLE 8.1

Advantages and disadvantages of slap & ship.

individual pallets to the staging area. Figure 8.2 illustrates the sequence of operation:

1. Cases are de-palletized.

2. The bar code for each unique case type is scanned. Product is identified and sorted for labeling and shipping requirements.

3. The number of cases is entered into a computer. A series of unique EPC codes are generated and stored in the local database.

4. Smart labels for each case are encoded and printed with unique EPC codes. The smart label RFID printer can be positioned within easy reach of workers by putting it on a mobile platform with a wireless network link. It can also be connected directly to the computer by Ethernet or by parallel interface.

5. Labels are applied to a pre-determined position on each case.

6. Once all cases have been labeled, the cases are re-palletized.

7. A pallet label is then encoded, printed and manually applied.

8. As a final check the pallet is read and compared with the pallet's original pick list.

9. At the end of each day management reports can be printed.

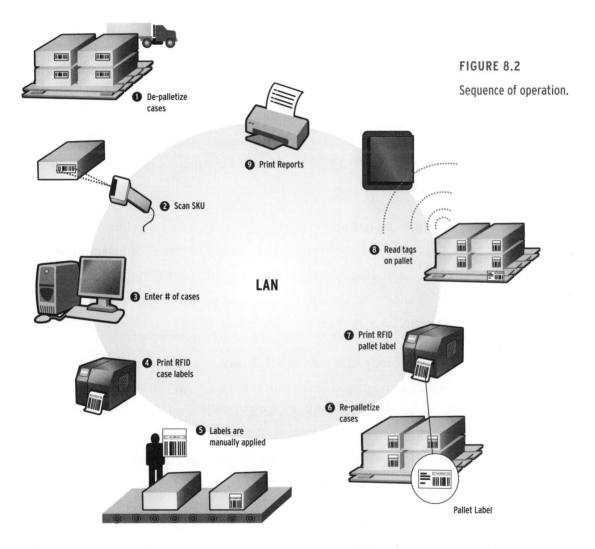

FIGURE 8.2

Sequence of operation.

1. Pallets of cases are de-palletized manually

2. The barcode for each unique case type is scanned

3. The number of cases are entered into the PC (A series of unique EPC codes are generated and stored in the local database)

4. RFID labels for each case are encoded and printed with unique EPC codes

5. Each is manually applied to a predetermined position on the case

6. Once all cases have been labeled, the cases are re-palletized

7. A pallet label is then encoded and printed and manually applied

8. As a final check the pallet is read and compared with the pallet's original pick list

9. At the end of each day management reports can be printed

Meeting Compliance Mandates with Slap & Ship

If you look at your RFID adoption curve as a conservative 7-10 year timeline, a slap & ship approach may be in use in various parts of your operation for a number of years. The time to transition from slap & ship to a more automated approach depends on a number of factors:

Product portfolio – High value items subject to lots of shrinkage justify the investment in RFID, and are immediate candidates for an automated approach. Products and packaging that are not "RF friendly" and require package and process engineering to achieve compliance may take longer to transition away from slap & ship.

Technology maturity & reliability – Automation requires a heavy initial investment and a longer period to re-coup costs. Automation usually entails building more rigidity into a process in order to achieve reliability. Since RFID standards and technology are continuing to evolve and change, there is risk in committing to a more automated approach too soon in the RFID adoption curve.

Competitive landscape – What your competitors do to meet RFID mandates will affect your decision when to automate. It will be important to acquire internal as well as a broad industry understanding of best practices. You should be developing a business case for automation well in advance of having to react to a competitor.

The tipping point, when you automate the application of RFID tags, is different for every company. The point can only be determined when you've evaluated the merits of slap & ship and print & apply approaches at various case volumes and points on the adoption curve.

SMART LABEL PRINT & APPLY APPROACHES

Print & apply is the term used for a semi- or fully-automated labeling process. Smart label print & apply requires a label applicator integrated with an RFID printer which can be synchronized with cases moving on a conveyor (Fig. 8.3).

→ SEE FIG. 8.3

Compliance Print & Apply

The application of a smart label is done at the point where the process knows that the pallet is destined for Wal-Mart (or another customer having an RFID initiative). Compliance print & apply can be done at a DC or outsourced to a third party logistics provider. A rationale for this approach is that a company wants to minimize the impact to its total operation since only a fraction of product goes to that customer.

Figure 8.4 illustrates a compliance print & apply approach. The majority of semi-automated approaches would use a conveyor system in conjunction with a label applicator integrated with a smart label printer. The conveyor is fed manually with cases from a de-palletizing

→ SEE FIG. 8.4

station. A separate smart label printer is stationed at the end of the line. Once all cases are tagged, they are manually re-palletized and shrink-wrapped. A pallet label is encoded and applied before the pallet exits through an RFID portal.

The benefits of this approach include:
• No interruption of shipments to other retail customers
• Controlled investments that support preparation for future needs

FIGURE 8.3

Label applicator with integrated RFID printer.

• Focus on identifying and solving RFID technical challenges.

• Methods would be applicable to other compliance requirements

• Integration with existing WMS and ERP

• Supports mixed pallet shipments

• Fully EPC compliant

FIGURE 8.4 Compliance print & apply system.

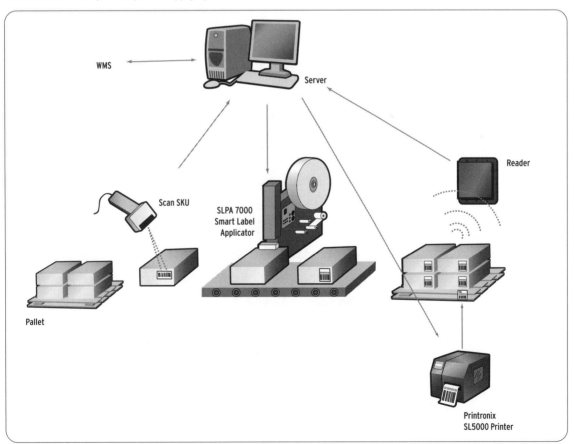

In-line RFID Print & Apply

To date, smart label print & apply technology is not yet capable of the 60-100 case per minute line speeds currently seen for some non-RFID print & apply units. Only high-volume production lines operate at 60-100 cases per minute, however. The normal is 20 to 30 cases per minute with a line speed of 60 feet per minute. Current smart label in-line print & apply systems provide reliable RFID encoding, validation and error recovery, with minimal operator intervention, at slightly lower line speeds. Companies may consider splitting lines and running parallel operations to increase output.

Smart label print & apply units can interface with practically any conveyor system via both software and hardware logic controllers. These logic controllers, through an input/output module, sequences the smart labeling process with upstream readers, conveyors, light stacks, alarms, and other equipment. There are two types of print & apply designs: loose loop and next-off. The loose loop mounts the printer on a standard label applicator. The printing and applying occur at two separate locations. A queue of printed labels – normally 10 to 20 labels depending on label size – exists between the printer and applicator. A next-off label printer applicator – which is what is normally referred to as a label printer applicator – is an integrated piece of equipment with no printed label queue – the printed label is immediately available to apply. The label printer applicator is preferred when the label data changes randomly or with each label.

Figure 8.4 shows an example of an integrated application. Error recovery is built into this approach. If a quiet tag happens to be on the smart label roll, it is detected automatically. The Printronix SLPA 7000e for example, keeps the label with the quiet or bad tag on the label liner and automatically sequences to a good tag. No placing in holding area required - which is an extra operation. The bad tags are rewound on the liner thereby facilitating collection and removal. A record is kept of the number of quiet and no-read tags. The applicator is instructed to place the bad label in a holding area, then applies the next label which is encoded and printed with duplicate information.

⬅ **SEE PAGE 155**

FIG. 8.4

SMART LABEL VALIDATION

The printing and encoding of smart labels needs to be a completely error-free operation. The smart label printer needs to recognize the occasional faulty RFID tag and prevent it from being used on a package. It also needs to provide validation data to a host computer for traceability.

A typical high-volume distribution center may print and scan 50,000 to 100,00 routing labels a day. An error rate of 3% can cost more than $1 million a year in such a circumstance. Since Wal-Mart is requiring 100% read rates for smart labels, and proof of validation, a smart label printer with EPC data validation plays an important part.

Validation Management – Integrated with a host computer, a smart label printer operates a closed loop system to validate its operation. The printer analyzes the incoming data stream and for each label associates the tag programming commands. The smart label has the data correlated against this data stream and validated against test parameters. If the label meets specifications, it is ready for use. If the label does not meet specifications, the label is voided, marked "bad," and a replacement label is printed and validated.

The validation activity is managed at the printer without host computer intervention. What is passed to the host is a validation log, indicating the number of labels printed, number canceled, and statistics by type of failure. By capturing this data and making it available to the host computer, the RFID printer closes the loop so that only known good smart labels enter the case/pallet labeling process.

A Pragmatic Path to Automation

A pragmatic approach requires an assessment of your product line using these rough categories:

RF friendly – Begin applying automation to SKU's with the highest internal ROI and the least sensitive content. RF friendly products are those that do not have high absorption or reflective qualities (liquids and metals). If they are packaged in paperboard material,

and are flexible to tag choice and location. A case analysis should indicate that you can get good reads regardless of orientation, even when loaded on pallets.

RF neutral – Cases that have dense product composition will exhibit moderate sensitivity to tag choice and location. They may be good picks because of their high value compared to labeling cost, and relative ease of labeling.

RF sensitive – Cases containing liquids and metals, or products in foil lined packages are RF sensitive. Air spaces between product can make a case somewhat RF lucent. Nonetheless, a case analysis will probably indicate labeling is consistent only within tight tolerances, with less-dense pallet loads. RF sensitivity may contribute to higher labeling costs as you move upstream to automate the process.

RF hostile – Cases that are opaque because they are densely packed, and contain products that attenuate the RF signal, will probably be the most expensive to label, and may be the biggest challenge to achieving ROI by applying upstream automation.

A business assessment should follow the product assessment prioritization. A decision to go forward should include an evaluation of available technology, system integration and process transformation requirements.

Sources and Further Reference

Section on "In-line RFID Print & Apply" contains numerous contributions by Rick Fox, President, Fox IV Technologies, Inc.

CHAPTER 9

Smart Labeling Systems Integration

The major benefits of RFID come from wireless computer-to-tag communication, but only if the data exchange is made useful. Through RFID, computers can identify, track, document and direct the movement of items through the supply chain with no human intervention. Data integration is a bottom up exercise. The groundwork involves the smart labeling of cases and pallets, and installing readers at key touch points in the delivery cycle. The next step is the integration of RFID with control and information systems.

OPERATIONAL DATA FLOW

Figure 9.1 illustrates the flow of RFID data. This model greatly simplifies what can be a complex layer of software, computer, storage, network and control systems:

Corporate planning – At the top of the information model are the human, financial, procurement, customer relations and product development resources of a company. The corporate planning system would be the keeper of the product catalog and ordering system, which includes product order numbers, SKUs, trade item identification, customer information and delivery requirements.

Manufacturing and distribution – A master schedule sequences product manufacturing, packaging and shipping. A forecasting system may drive case builds and inventory in anticipation of receiving the actual customer order. Product and order information

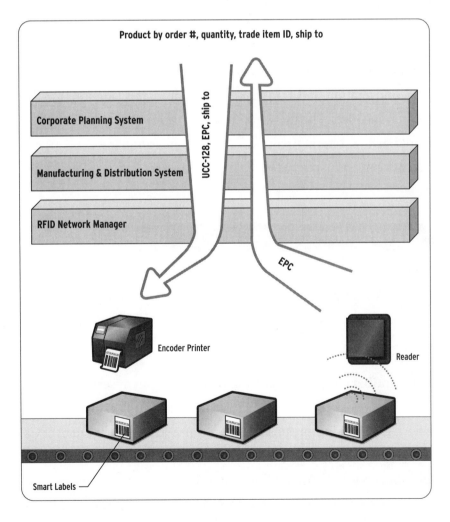

FIGURE 9.1

Operational data flow.

converge at the point where the actual labeling and order fulfillment takes place. This may be at the end of a packaging line or at a warehousing and distribution center.

RFID network – The RFID network starts by managing the labeling of cases under the direction of the manufacturing and distribution system. EPC numbers may be created here, or passed down from another system. EPC data is combined with bar code and other label information and translated into a command language for the printer/encoder. The network manager also manages the RFID reader arrays, filtering redundant reads and compiling lists of tags. The network manager passes tag lists to the other layers of the system.

DEPENDENCIES AND POINTS OF CONTROL

Whether the smart labeling system is minimally designed for initial compliance requirements or has a high level of integration with other systems, a number of areas need to be managed. Critical dependencies and points of control include:

EPC numbering – Case, pallet and eventually item serial numbers need to be created, encoded in EPC format, and recorded somewhere for tracking purposes. You can use a PC to generate EPCs along with other labeling information, and locate it right next to a labeling station. You may find this approach to be the easiest and least expensive for near term compliance driven applications. Label application software companies are now coming out with EPC modules to handle the task. Alternatively, a software module within your supply chain execution system may generate an EPC and embed it in the print job stream. You can also purchase pre-numbered RFID tags. In this situation you would need to set up a system at point of application to create a database that matches GTIN with EPC.

Trade item data – Serialized trade items in EPC format should link with your master catalog. Options include converting the catalog to a new database schema and storing the new information there, or using a separate software module that does a look up and cross-reference between your master catalog and assigned and available EPC numbers.

Process management – Your distribution system needs to poll readers for tag data, validate the information on packages and pallets, and use that information to trigger next steps in the process.

Exception handling – The process must react to quiet tags, mis-labeled cases, and trigger error recovery routines, alerts and management reports.

ASN – Advanced shipping notice information requires reading of a pallet EPC, correlation of case EPC numbers with the pallet, and EDI transmission of the shipping confirmation.

← SEE PAGE 93
FIG. 5.2

Management reporting – Records need to be kept to validate the tag encoding, smart labeling and reading process. They become part of the product chain of custody.

Returns – EPC number lookups and handling systems for returns may require modifications to existing systems. Tag readers help automate these systems to reduce special handling labor.

Middleware – RFID systems create a lot of data. Software components, called middleware, are used to filter, store and forward the data to other systems, thus reducing network traffic. Commercial application integration software can package EPC data with descriptions in PML or some other application-neutral format so it can be made available to supply chain partners using a publish and subscribe protocol.

ENTERPRISE-WIDE SMART LABEL PRINTING MANAGEMENT

Enterprise integration of smart label printers offers another level of capability and control. With print management software, all label printing operations can be monitored from a web browser practically anywhere on the planet. Printers can be made visible to ERP and systems management software applications. Alarm conditions can be set and monitored, diagnostics run, and alerts sent to the appropriate individuals. In addition, remote print management allows centralized configuration control, lockouts of printer configurations, and coordinated updates to firmware. See Figure 9.2.

GLOBAL EXCHANGE OF SUPPLY CHAIN DATA

Electronic product codes are the lowest-level link in a multi-tier information model conceived by the Auto-ID Center (now

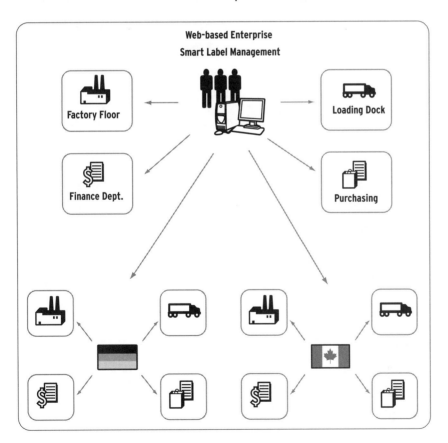

FIGURE 9.2

Enterprise smart label print management system.

EPCglobal). As envisioned by EPCglobal, EPCs are assigned, catalogued, tracked and managed within a collaborative system associated with the Internet. Using an EPC, a host computer can look up via the Internet stored information about a specific item, including the manufacturer, product classification, handling, use and status in the supply chain.

EPCglobal, in partnership with VeriSign, the company that manages the Internet Domain Name Service (DNS), has begun laying the groundwork for what is called the EPC Network. The EPC Network will leverage the Internet to support the cataloging of EPCs for supply chain partners. The overall information management concept and system architecture includes these other system components:

Object Name Service (ONS) – Like the Internet Domain Name Service (DNS), which provides lookups for locations on the World Wide Web, ONS serves as a registry for distributed EPC Information Services databases. Managed by VeriSign, ONS will link an EPC with the IP address of a database that stores relevant information.

Information Services – EPC Information Services are the actual data repositories used to store unique item data. These are distributed databases maintained by companies, and referenced through ONS, like how DNS points to web sites on the Internet.

EPC Discovery Service – The directory service stores EPC history. It serves as a chain of custody service, providing tracing information as a product moves through the supply chain.

Physical Markup Language (PML) – PML provides a common reference language for electronic communications amongst trading partners. Like HTML, the web page description language, and eXtensible Markup Language (XML), which is a generalized data description language, PML will extend EPC to include associated information of value to the supply chain. PML pages for each EPC can be set up, maintained and shared by the EPC product manufacturer or "owner." PML descriptors might include the following kinds of information:

• Expiry dates and safe handling instructions

• Ingredients and composition

• Physical properties, including telemetry information (where it is located at any moment in time).

• Procedural information, such as which processing, packaging and quality control steps have occurred.

Data flow is depicted in Figure 9.3. Middleware intercepts the massive amount of tag data available from local readers and rationalizes it into information for use with Warehouse Management Systems (WMS), Enterprise Resource Planning (ERP) systems and elsewhere. Without such a system, networks and computers would collapse from the RFID traffic generated by tags going through the

→ SEE FIG. 9.3

manufacturing and packaging process.

EPC data architecture not only allows for faster identification of product, but supports the movement of information through the value chain and product life cycle. Ultimately this will lead to business efficiencies. An associated standardization effort, called Global Data Synchronization (GDS), is moving toward the establishment of harmonized item data.

FIGURE 9.3

EPC information flow.

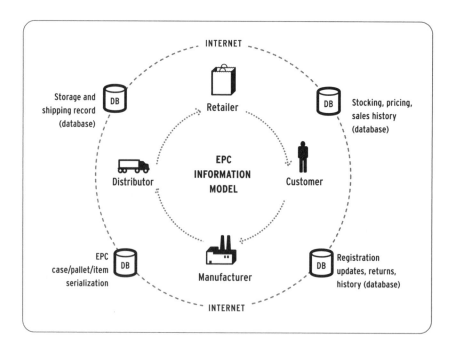

GLOBAL DATA SYNCHRONIZATION

Global data synchronization, or GDS, is an industry-wide initiative to create standardized forms for product information that can be shared electronically by manufacturers, distributors and retailers. UCCnet facilitates the initiative. UCCnet provides services for synchronizing product data and achieving compliance. In the retail industry, mandates from Wal-Mart, Lowes and others are driving suppliers to register with UCCnet.

Global data synchronization involves a number of internal and external activities (see Figure 9.4):

→ SEE FIG. 9.4

Data cleansing – For items that it produces and trades, a supplier must adopt standard names, descriptions and identities. This activity can take the form of an internal conversion of master data files, or an application that manages the correlation of internal data files and UCCnet compliant formats. Data cleansing can be a considerable effort for a company, especially when their internal systems (accounting, manufacturing, distribution, etc.) do not represent item data consistently.

UCCnet registration and synchronization – After data cleansing, a company registers its trade items with UCCnet. This neutral data repository gives trading partners access to the item data, including product numbers and pricing. Automatic updates are used to ensure that the UCCnet registry reflects current item data for the company.

The elimination of error and the gaining of transaction efficiencies are the end goals of global data synchronization. One study reports that 60% of all invoices generated in the consumer goods industry have errors, with an estimated cost to correct them at $40 to $400 per incident. The monetary benefits to broad industry adoption of

FIGURE 9.4

GDS process.

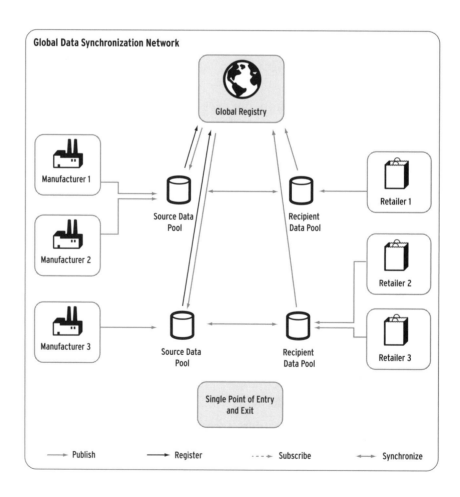

GDS could translate to $1 million saved for each $1 billion in sales, according to a study by AT Kearney and Kurt Salmon Associates.

Although a number of implementation issues related to GDS are yet to be resolved, the basic framework is in place for North America, as are the industry mandates for its adoption. Third party solutions and services for data cleansing and UCCnet compliance work are available.

Adoption of EPC and GDS go hand in hand. They are foundational capabilities, as shown in Figure 9.5. While RFID technology can provide more detailed and faster tracking of product in the supply

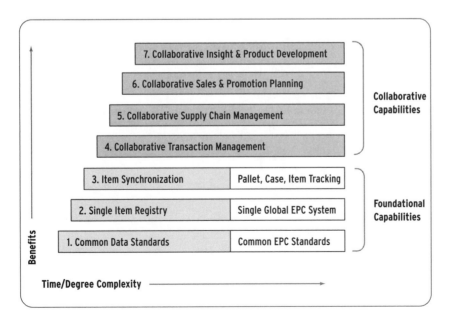

FIGURE 9.5

EPC and GDS are foundation activities for supply chains.

chain, it is not useful unless the exchange of data is correct. A data cleansing activity of some form is required when a company adopts EPC, just as its necessary to global data synchronization. Otherwise, a company is at the risk of simply automating a "garbage in, garbage out" system.

Sources and Further Reference

Meeting the Retail RFID Mandate, white paper, A.T. Kearney, November 2003.

Connect the Dots, white paper, A.T. Kearney and Kurt Salmon Associates, February 2004.

The EPC Network: Enhancing the Supply Chain, white paper, VeriSign, 2004.

Contributions by Rick Fox, President, Fox IV Technologies, Inc.

RFiD
(((SMART)))™

Products Section

PRINTRONIX®

www.printronix.com

SMART LABEL DEVELOPER'S KIT

The Printronix Smart Label Developer's Kit is specifically designed to help you fast track an RFID pilot. The kit creates a complete smart label developer's environment, something that cannot be done with an "off-the-shelf" printer.

Transition from Bar Codes to Smart Labels

Printronix's unique RFID Software Migration Tools will allow you to effortlessly move from the printing of bar code labels to encoding smart labels, putting your pilot program on fast track. The tools include a suite of applications that convert standard UPC and GTIN print streams into printer commands, which encode UPC or GTIN on the smart labels, in addition to printing the standard bar code label. Simply select from one of our many migration tools through the printer control panel. The printer will automatically enable the encoding of smart labels without the need to change your print stream, application or environment.

EPC Compliant Labels

When you are ready to advance your RFID pilot to the encoding of EPC labels, the Developers Kit will grow with you. Our Software Migration Tools include an EPC tool built into the printer, allowing you to use any existing label design software to create an EPC RFID tag. The Printronix Smart Label Developer's Kit also comes with a complete set of programming manuals. Use the step-by-step instructions to add EPC commands to your existing PGL® or other printer data streams.

The kit contains everything necessary to begin the encoding of smart labels:

- SL5000^e MP-multi-protocol RFID thermal printer. A web-enabled industrial-grade thermal bar code printer designed for exacting label applications.
- Integrated RFID UHF encoder
- 1,000 certified RFID smart labels
- One premium wax thermal ribbon
- PrintNet® Enterprise Ethernet connectivity
- Application notes
- Reference notes
- Technical support
- Programming manuals
- Software migration tools that permit the seamless encoding of smart labels.

The kit also includes PrintNet® Enterprise, the web-enabled remote network print management system that provides instantaneous visibility to every network printer. It allows users to simultaneously configure and efficiently manage an unlimited number of Printronix printers. This edition also supports management of RFID encoder capabilities.

SMART LABEL PRINTERS FOR RFID PILOTS

As the adoption of RFID technology accelerates in retail and government supply chain operations, so must the printing solutions for smart labels. SmartLine™ RFID printers enable encoding and printing of various label sizes and antenna designs that have emerged as popular standards through early adopter pilot programs. In addition to label size and antenna design, the need for specific encoding technology to support today's EPC standards with an easy migration path to tomorrow continues to be important.

SL5000^e™ RFID PRINTERS

Two models provide customers with a choice of encoding technologies:

- SL5000^e MP* – provides multi-protocol capabilities to address applications that require Class 0, 0$^+$, 1 standard.

- SL5000^e C1 – provides capabilities to address applications that require an optimized Class 1 standard solution.

KEY PRODUCT ATTRIBUTES OF THE SL5000^e

- Integrated RFID UHF 915MHz encoder.

- Smart Label encoding validation and overstrike capability.

- PGL and ZPL™ programming languages with RFID command sets.

- Unique dual-motor driven ribbon system eliminates clutch replacement and greatly reduces the risk of ribbon wrinkle for superior print quality.

- Exclusive snap-in print head reduces service calls by allowing operator to easily change print heads and change from 203DPI to 300DPI printing without firmware or software changes.

- Wireless/Ethernet option provides real-time data access and local printing flexibility; includes Printronix PrintNet® Enterprise, a

web-enabled remote network print management system that provides instantaneous
visibility to every network printer and allows
users to simultaneously configure and efficiently manage an unlimited number of
Printronix printers; this edition of
PrintNet Enterprise also supports
management of the additional RFID
encoder capabilities.

SMARTLINE PRINTER SPECIFICATIONS

SL5000^e RFID PRINTERS

SL5000^e MP	Multi-protocol – AWID Technology
	Support for Class 0, 0$^+$, 1
	915 MHz UHF encoder
SL5000^e C1	Class 1 protocol – Alien Technology
	Support for Class 1, 915 MHz UHF encoder

PRINTING CHARACTERISTICS

Print Speed	SL5204^e: 10 IPS @ 203DPI (254mm/sec)
	SL5304^e: 8 IPS @ 300DPI (203mm/sec)
Printing Methods	Thermal Transfer or Direct Thermal
Resolution	203/300 DPI (operator interchangeable)
Printable Width	4.1" max

RECOMMENDED LABEL CHARACTERISTICS

RFID Inlay Labels:

Alien 'Squiggle' Class 1	4" x 2", 4" x 4", 4" x 6", 4" x 8"
Alien 'M-tag' Class 1	4" x 4", 4" x 6", 4" x 8"
Rafsec #313 'Psychedelic' Class 1	3" x 3"
Matrics 'Dual Dipole' Class 0, 0$^+$	3" x 3" (via software upgrade)

RFID ENCODING MODES

Integrated RFID reader and antenna assembly

Supports Class 0, 0$^+$, 1 tags

Operation Modes	Write/Verify/Print – writes RFID data to tag and verifies contents are written correctly, while also printing the desired image
Error Handling Modes	Overstrike – when a bad RFID tag is detected, overstrikes label and applies the data with the next label
	Stop – when a bad tag is detected, stops the printer to allow for user invention
Statistics Tracking	Tracks number of tags written to and number of bad tags detected

RFID SOFTWARE MIGRATION TOOLS

Migration tools allow RFID tags to be automatically programmed with data from the following bar code symbologies:

GTIN carriers ITF-14 and UCC/EAN-128

EPC data within a Code 39

UPC-A, EAN-8, EAN-13

Some restrictions may apply. Contact a Printronix Certified RFID Integrator for details.

MEDIA HANDLING CHARACTERISTICS

Tear-Off	Individual label tear-off
Tear-Off Strip	Label strips tear-off
Continuous	Labels print continuously
Peel	Labels peel from liner without assistance (peel mode requires rewind option)

MEDIA COMPATIBILITY

Media Types	Roll or fanfold
	Die cut or continuous
	Labels, tags and tickets
	Paper, film or synthetic stock
	Thermal Transfer or Direct Thermal
Media Width	1.0" to 4.5" (SL5204^e/SL5304^e)
Media Thickness	0.0025" to 0.010"
Roll Core Diameter	3.0" (7.6 cm)
Maximum Roll Diameter	8.0" (20.9 cm)
Thermal Transfer Ribbon	
Ribbon Width Range	1.0" to 4.33" (SL5204^e/SL5304^e)
Ribbon Capacity	625m

RECOMMENDED RIBBON CHARACTERISTICS

Ribbon	625m length
	4.33" width
	Model 8500 premium wax formulation
	Model 8300 wax formulation

OPERATOR CONTROLS & INDICATORS

Operator Controls	Off Line-On Line, Test Print, Job Select, Form Feed Menu, Cancel, Enter
Message Display	32 character
Indicators	Off Line-On Line, Menu

PROGRAMMING LANGUAGE

Standard	PGL – Printronix Graphics Language
	ZPL – Zebra Programming Language

BARCODE SYMBOLOGIES AVAILABLE

Code 39, Code 128 (A, B, C) Codabar, Interleaved 2 of 5, FIM UPC-A, UPC-E, UPC-EO, EAN 8, EAN 13, Code 93, Postnet, Postbar UCC/EAN 128, PDF 417, UPS Maxicode, Royal Mail, Datamatrix

SENSING METHODS

Transmissive, Reflective (Gap, Mark, Notch, Continuous Sensing Form)

INTERFACES

Standard	Serial RS232
	IEEE 1284 (Centronics)
Optional	Ethernet (PrintNet)
	Dual Ethernet/Wireless (802.11b)

FONTS, GRAPHICS SUPPORT, WINDOWS DRIVERS

Fonts	OCRA, OCRB, Courier, Letter Gothic CG
	Triumvirate Bold Condensed
Graphic Support	PCX and TIFF file formats
Windows Drivers	Windows 95/98/2000/XP

MEMORY

DRAM	8Mb standard (16Mb optional)
Flash	4Mb standard (10Mb optional)

POWER REQUIREMENTS

Line Input	90–264 VAC (48–62Hz), PFC
Power Consumption	150 watts (typical)
Regulatory Compliance	FCC-B, UL, Energy Star mode (<45 watt)

ENVIRONMENTAL CONSIDERATIONS

Operating Temperature	5°C to 40°C
Dimensions	11.7" W x 20.5" L x 13.0" H (SL5204^e/SL5304^e)
Printer/Shipping Weight	42lbs/51lbs (SL5204^e/SL5304^e)

SMART LABEL PRINT AND APPLY SYSTEM

The SLPA™ 7000 is a smart label printer-applicator solution that encodes, prints and applies—all in one unit. Combining RFID smart label printing technology with applicator capability, this RFID smart label solution delivers fast, accurate, cost-effective encoding and printing to users with site-specific requirements.

- Enables supply chain operations using encode, print-and-apply label applicators to automate the application of RFID smart labels.

- A fully integrated RFID printer and applicator is designed for heavy industry environments.

- Supported orientations include top and side label application and roll-on/front/back applications.

- Built-in quality control features identify and reject "bad" or "quiet" labels—ensure 100% high-performance RFID smart labels every time.

- Supports multiple label sizes providing flexibility to address different application requirements—reducing the total quantity of print-and-apply label applicators required.

RFID SMART LABELS

Printronix Smart Labels combine the technology of thermal transfer pressure sensitive labels with Radio Frequency Identification (RFID). The pressure sensitive label has an embedded RFID tag. The RFID tag is a microchip attached to an antenna and uses a WORM technology. These tags are capable of storing up to 64-bits of information and will advance to handle 96-bits as RFID evolves.

Under software control, the Printronix Printer prints on the label and writes data to the RFID tag. Once written, the Printronix system then validates that the EPC information was written to the RFID tag properly.

Designed for Emerging Standards

RFID standards are continuing to evolve. For distribution center applications, the Ultra High Frequency (UHF) of 915 MHz is emerging as the standard frequency for reading pallets and boxes of goods as they pass into a warehouse, distribution center, or factory floor. This frequency provides increased range and faster data transfer rates than lower frequencies. Printronix Smart Labels support 915 MHz, Class 1 communications protocol and will progress to support emerging classes and frequencies.

Programming Through PGL

RFID commands have been added to the Printronix PGL language, the industry standard bar coding graphics language, to provide easy transition from bar codes to Smart Labels. This allows Printronix RFID solutions to be easily integrated into your current label printing applications at a lower cost than other RFID hardware providers.

Migration of Existing Applications

Printronix has developed a set of Software Migration Tools (SMT) to enable the printer to automatically create RFID commands from your existing UCC label bar code application. The tool will encode bar code data in your current data stream to the Printronix RFID Smart Label tag.

The SL5000e MP and SL5000e C1 support the following label sizes with various antenna designs.

Alien 'Squiggle' Class 1	4" x 2", 4" x 4", 4" x 6", 4" x 8"
Alien 'M-tag' Class 1	4" x 4", 4" x 6", 4" x 8"
Rafsec #313 'Psychedelic' Class 1	3" x 3"
Matrics 'Dual Dipole' Class 0, 0+ **	3" x 3"

* Class 0 and 0+ capabilities will be available via a software upgrade, please contact Printronix or a Printronix Certified RFID Integrator for availability.

** SL5000^e MP only

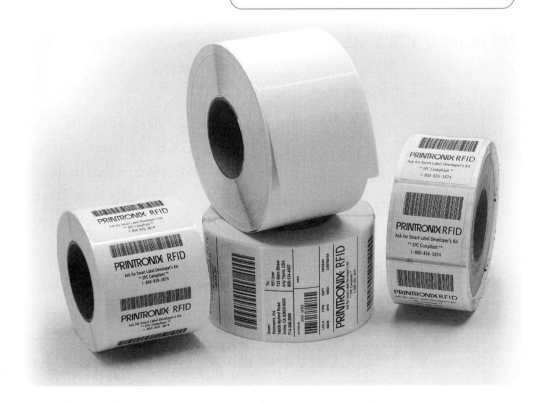

THERMAL BAR CODE PRINTERS

The T5000^e SR thermal printers, are web-enabled, industrial-grade thermal bar code printers specifically configured for easy field upgradeability to RFID encoding. These products have been designed to operate at 24/7 duty cycles in manufacturing or distribution environments and provide high quality printing performance.

KEY PRODUCT ATTRIBUTES OF THE T5000^e SR

- Aluminum die-cast design dampens vibration and maintains precise printer alignment.

- Unique dual-motor driven ribbon system eliminates clutch replacement and greatly reduces the risk of ribbon wrinkle for superior print quality.

- Cantilever head mechanism design makes it easy to side-load, clean and maintain.

- Exclusive snap-in print head reduces service calls by allowing operator to easily change print heads and change from 203DPI to 300DPI printing without firmware or software changes.

- Ventless system operates in environments with airborne particulate matter without compromising performance.

- Wireless/Ethernet option provides real-time data access and local printing flexibility.

SL5000^e RFID UPGRADE KITS

Choose from two different upgrate kits for RFID encoding. These upgrade kits have everything needed to upgrade at T5000^e SR thermal printer into either a multi-prototcol or EPC class 1 optimized SmartLine™ printer.

SL5000^e™ MP UPGRADE KIT –
MULTI-PROTOCOL TECHNOLOGY

- 915MHz UHF encoder (AWID Technology)
- Support for Class 0, 0$^+$, 1
- RFID software program kit
- All necessary cables and connectors
- PGL™ RFID programming manuals
- Software migration tools for seamless encoding of smart labels

SL5000^e C1 UPGRADE KIT –
CLASS 1 PROTOCOL TECHNOLOGY

- 915MHz UHF encoder (Alien Technology)
- Support for Class 1
- RFID software program kit
- All necessary cables and connectors
- PGL™ RFID programming manuals
- Software migration tools for seamless encoding of smart labels

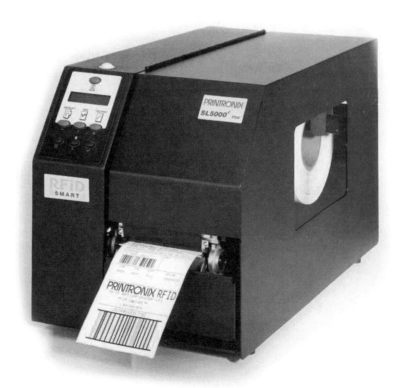

SMARTLINE PRINTER SPECIFICATIONS

SL5000^e UPGRADE KITS

SL5000^e MP	Multi-protocol – AWID Technology
	Support for Class 0, 0$^+$, 1
	915 MHz UHF encoder
SL5000^e C1	Class 1 protocol - Alien Technology
	Support for Class 1, 915 MHz UHF encoder

PRINTING CHARACTERISTICS

Print Speed	T5204^e SR: 10 IPS @ 203DPI (254mm/sec)
	T5304^e SR: 8 IPS @ 300DPI (203mm/sec)
Printing Methods	Thermal Transfer or Direct Thermal
Resolution	203/300 DPI (operator interchangeable)
Printable Width	4.1" max

RECOMMENDED LABEL CHARACTERISTICS

RFID Inlay Labels:

Alien 'Squiggle' Class 1	4" x 2", 4" x 4", 4" x 6", 4" x 8"
Alien 'M-tag' Class 1	4" x 4", 4" x 6", 4" x 8"
Rafsec #313 'Psychedelic' Class 1	3" x 3"
Matrics 'Dual Dipole' Class 0, 0$^+$	3" x 3" (via software upgrade)

RFID ENCODING MODES

Operation Modes	Write/Verify/Print – writes RFID data to tag and verifies contents are written correctly, while also printing the desired image
Error Handling Modes	Overstrike – when a bad RFID tag is detected, overstrikes label and applies the data with the next label
	Stop – when a bad tag is detected, stops the printer to allow for user invention
Statistics Tracking	Tracks number of tags written to and number of bad tags detected

RFID SOFTWARE MIGRATION TOOLS

Migration tools allow RFID tags to be automatically programmed with data from the following barcode symbologies:

GTIN carriers ITF-14 and UCC/EAN-128

EPC data within a Code 39

UPC-A, EAN-8, EAN-13

Some restrictions may apply. Contact a Printronix Certified RFID Integrator for details.

MEDIA HANDLING CHARACTERISTICS

Tear-Off	Individual label tear-off
Tear-Off	Strip Label strips tear-off
Continuous	Labels print continuously
Peel	Labels peel from liner without assistance (peel mode requires rewind option)

MEDIA COMPATIBILITY

Media Types	Roll or fanfold
	Die cut or continuous
	Labels, tags and tickets
	Paper, film or synthetic stock
	Thermal Transfer or Direct Thermal
Media Width	1.0" to 4.5" (T5204^e SR/T5304^e SR)
Media Thickness	0.0025" to 0.010"
Roll Core Diameter	3.0" (7.6 cm)
Maximum Roll Diameter	8.0" (20.9 cm)
Thermal Transfer Ribbon	
Ribbon Width Range	1.0" to 4.33" (T5204^e SR/T5304^e SR)
Ribbon Capacity	625m

RECOMMENDED RIBBON CHARACTERISTICS

Ribbon	625m length
	4.33" width
	Model 8500 premium wax formulation
	Model 8300 wax formulation

OPERATOR CONTROLS & INDICATORS

Operator Controls	Off Line-On Line, Test Print, Job Select, Form Feed Menu, Cancel, Enter
Message Display	32 character
Indicators	Off Line-On Line, Menu

PROGRAMMING LANGUAGE

Standard	PGL™ – Printronix Graphics Language
	ZPL™ – Zebra Programming Language

BARCODE SYMBOLOGIES AVAILABLE

Code 39, Code 128 (A, B, C) Codabar, Interleaved 2 of 5, FIM UPC-A, UPC-E, UPC-EO, EAN 8, EAN 13, Code 93, Postnet, Postbar UCC/EAN 128, PDF 417, UPS Maxicode, Royal Mail, Datamatrix

SENSING METHODS

Transmissive, Reflective (Gap, Mark, Notch, Continuous Sensing Form)

INTERFACES

Standard	Serial RS232
	IEEE 1284 (Centronics)
Optional	Ethernet (PrintNet)
	Dual Ethernet/Wireless (802.11b)

FONTS, GRAPHICS SUPPORT, WINDOWS DRIVERS

Fonts	OCRA, OCRB, Courier, Letter Gothic
	CG Triumvirate Bold Condensed
Graphic Support	PCX and TIFF file formats
Windows Drivers	Windows 95/98/2000/XP

MEMORY

DRAM	8Mb standard (16Mb optional)
Flash	4Mb standard (10Mb optional)

POWER REQUIREMENTS

Line Input	90–264 VAC (48–62Hz), PFC
Power Consumption	150 watts (typical)
Regulatory Compliance	FCC-B, UL, Energy Star mode (<45 watt)

ENVIRONMENTAL CONSIDERATIONS

Operating Temperature	5°C to 40°C
Dimensions	11.7" W x 20.5" L x 13.0" H (T5204^e SR/T5304^e SR)
Printer/Shipping Weight	42lbs/51lbs (T5204^e SR/T5304^e SR)

PrintNet ENTERPRISE

PrintNet® Enterprise is the only system that lets you manage all your print operations—whether it's a single warehouse or a global infrastructure —from one computer. This advanced Web-based management tool combines a fully integrated Ethernet adapter and Java-based remote management software to deliver unparalleled remote printer management adaptability, remote diagnostics and help desk tools. It supports SNMP, which allows you to use SNMP managers like HP Open View, Tivoli, Computer Associates Unicenter TNG, Sun Net Manager and Castle Rock SNMPc.

With PrintNet Enterprise you get:

- Instant RFID printer status, which indicates which printers are in use or idle and which need consumables or repairs.

- Instant alerts, which notify you via e-mail (or routed to your pager or cell phone) if an RFID printer situation requires immediate action.

- Instant control, which allows you to remotely change printer settings, update firmware, download fonts and lockout local operators from making changes.

- Instant RFID printer organization, which enables you to configure printers from a single action.

- Instant diagnostics, which remotely diagnoses problems via a Web browser.

Visibility

Whatever the status of your printers, you will receive a message with a clear and comprehensive view of your entire printer operation.

- **Main Status Screen** – Displays the status of every printer under management, including a "gas gauge" indicating ribbon levels and color-coded icons highlighting problem spots and text messages.

- **SNMP** – Fully supports the printer MIB, making PrintNet® Enterprise-equipped printers visible and accessible with tools such as HP OpenView, WebJet Admin, Tivoli, Computer Associates Unicenter TNG, Sun Net Manager, and Castle Rock SNMPc.

Instant Alerts

You are instantly notified when a printer situation demands immediate attention. Whether by e-mail, pager or cell phone, these instant messages can be generated to a specific responder, varying according to class of alert. Even if the printer is totally dead due to catastrophic power supply, controller failure, or loss of power, an instant alert is generated.

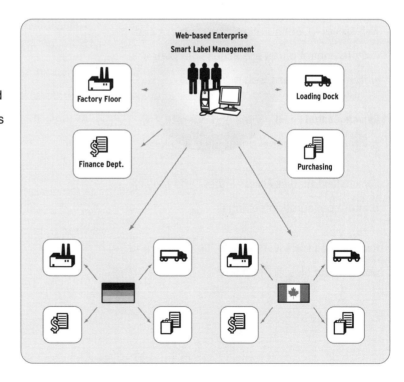

Alerts can be delivered in numerous ways

- Email Message
- SNMP Traps
- Syslog
- Logged to File

Remote Presence

From any network PC or workstation, you can quickly diagnose a problem without the need to travel to the printer.

- **Web Access** – Configuration management and remote control panel are accessible over the web using standard web browsers.

- **Remote Control Panel** – View and operate any printer's control panel online as if it is local.

- **Remote Configuration Access** – View and modify printer configuration settings.

- **Remote Flash File Access** – View, download, and upload user files in the printer.

Whatever the situation demands, you have total control of your printers. Change printer settings, update firmware, download fonts, clear buffers – even lock out local operators from making changes. All these controls are available to you without ever going to the printer.

- **Configuration Editor** – Creates, saves, and downloads configuration tables consisting of all parameters.

- **Configuration Locking** – Controls which printer settings, firmware updates and downloaded fonts can and cannot be modified by local operators, including the ability to completely exclude them from making any changes.

- **Flash File Manager** – Add, remove and move files (such as fonts and predefined forms) stored in the printer's flash file area.

- **Automated Firmware Updates** – Firmware versions of one or more printers can be updated at the same time. All configurations and files are automatically retrieved and restored as part of the update.

- **Speed Keys** – One-touch access to change the most commonly accessed printer settings.

- **Automatic Configuration Switching** – Tie a printer configuration to one of eight internal print queues. Whenever a job is sent to that queue, the specified configuration is automatically loaded. This feature ensures that your jobs will be processed with the correct configuration every time.

Centralized Programming

You can ensure all printers have a common configuration by programming once and sending the programming to all your printers.

- **Individual, Group or Broadcast Management** – Perform an operation to entire groups of printers with one set of commands. There's absolutely no need to configure printers individually.

- **Printer Database** – Allows organization of printers into groups matching applications, locations, or any other criterion the environment demands.

OPTIONS AND ACCESSORIES

Wireless RFID Printing

Today's just-in-time (JIT) manufacturing and logistics environments require advanced printers that can perform where and when they are needed. Although wireless LAN technology provides a viable solution, it also increases the need for true printer management to ensure consistency and diagnostic control.

Printronix SmartLine Mobile Print System (MPS) addresses this need. Combined with our RFID printer, MPS delivers smart labels directly to locations that need it the most. Its ergonomic design, durable steel frame and heavy-duty wheels combine maximum maneuverability with minimal maintenance to offer these benefits:

- Reduced centralized printing and manual distribution by printing smart labels on the spot.

- Scalable deployment for maximum printing flexibility at compliance locations.

- Elimination of expensive, time-consuming, hard-cabled system reconfigurations when changeovers occur.

- Centralized control, consistency and diagnostics previously available only in hard-cabled managed systems.

- Optimized cost efficiency and productivity.

- Higher return on investment (ROI) due to accelerated data and inventory movement.

Once installed in a wireless network, and powered by PrintNet Enterprise, this unique technology allows printing of real-time data in any location. It eliminates time-consuming and expensive re-configuration of hard-cabled systems when warehouse or manufacturing floor changeovers occur. It also provides centralized control, consistency and diagnostics previously available only in hard-cabled PrintNet Enterprise-managed systems.

Integration into Legacy and Proprietary Language Environments

Printronix provides tools that enable the T5000^e to integrate seamlessly into virtually any proprietary language environment. This enables greater bar code labeling efficiency, and ultimately, a link to higher performance and reliability in the delivery cycle of your products. Printronix Printer Interpreter (PPI) integrates T5000^e printers in Zebra, Intermec and TEC installations easily. Once integrated, they maintain all capabilities used in a Printronix environment.

Mobile Print System

Bar coding applications can be put into place anywhere, any time. Combining the wireless T5000^e with a rugged industrial cart delivers thermal bar coding directly to locations that need it most. The system includes an ergonomically designed cart with a durable steel frame and heavy-duty wheels for maximum maneuverability and minimal maintenance. Critical features such as battery-life warning are integrated with PrintNet Enterprise and fully visible to operators.

TN5250/3270 Ethernet Option

As the market moves toward 100 Base-T and 1 Gigabit Ethernet, maintaining an out-dated Twinax or Coax network becomes increasingly difficult to justify. Two power factors impair a migration: legacy applications must be changed

and a high level of print job control must be maintained. The TN5250/3270 Ethernet Option maintains the job control and visibility of Twinax or Coax networks, yet combines it with the sped and efficiency of Ethernet. In addition, the option provides:

- Automatic confirmation of which pages have and have not printed.
- Knowledge of exactly where an error has occurred in batch printing.
- Ability to restart printing on only those pages needed.
- Full access to all the features and functionality of PrintNet Enterprise.

On-Line Data Validation (ODV™)

Printronix Online Data Validation (ODV) is an option exclusive to our T5000^e thermal printer series. It's a tool that can save time, money, stalled productivity… and a lot of frustration. This unique quality control option can inspect every bar code on every label printed on a Printronix T5000^e. Specifically, ODV technology analyzes each bar code just after it is printed. This 100% analysis verifies that the print image falls well within each bar code's published quality specifications to ensure the bar code will scan successfully. If any code on a label fails, the label is automatically cancelled and a replacement label is printed.

Today's supply chains depend on Automatic Identification and Data Capture (AIDC) technology for proper routing and distribution. AIDC allows no tolerance for bar code scanning failure while products pass through the supply chain. Bar codes must read 100% of the time. Any reading failure can escalate into a rejected item, manual recovery steps, time delay, a loss in productivity, an unhappy customer, compliance failure fines, or a even a loss of business.

Printronix created Online Data Validation to help you avoid such problems. In doing so, ODV:

- Evaluates every bar code on every label to provide total scanning assurance.
- Automatically cancels and reprints any label on which a bar code failed.
- Eliminates the need for costly human inter-

vention in the bar code validation process.

- Combined with Printronix PrintNet® Enterprise, ODV closes the loop on bar code quality control and provides unprecedented visibility of mission-critical print operations.

ODV Data Manager

Imagine a world in which bar code verification systems could integrate with enterprise networks, eliminating costly bar code rejection and capturing real-time data about printing applications. Imagine if customers could monitor and review all bar codes printed and then merge the data into custom reports, all while improving Return on Investment. With our exclusive ODV Data Manager integrated with PrintNet Enterprise, imagination is not needed. ODV Data Manager provides robust data capturing and exporting capabilities and the ability to evaluate data within each bar code. It merges information into databases and ports it to any application, such as SAP or Oracle.

ODV Data Manager helps customers manage industrial network printer systems, while providing a safeguard against the high costs associated with bar code scanning and data accuracy failure.

The Printronix Online Data Manager provides the following functions:

- Ability to capture bar code and validation data.
- Ability to export captured data to either an ASCII file or SQL database.
- Ability to stop the device and alert the operator of a problem.

All data captured by the ODV Data Manager is date and time stamped, and associated with the printer generating the data. Collected data is either stored in a history file or immediately exported to an external SQL database. Data stored in the history file can be saved in either a comma separate variable (CSV) file format, or as an XML report. Data from a single printer or multiple printers can be saved in either a single file or in multiple files.

GENERAL PURPOSE INPUT OUTPUT MODULE

Printronix has expanded its growing portfolio of printer management solutions and enhanced the functionality of its thermal printers with the General-Purpose I/O (GPIO) Module and GPIO Manager. The GPIO accessory module is designed specifically for those who want to integrate Printronix T5000^e thermal printers with other systems, such as label application systems, programmable logic controllers, light stacks and host computers.

The GPIO accessory module is comprised of an I/O circuit board that is easily installed in a T5000^e printer. Simple printer menus allow programming of specific interface signals for proper polarity or logic functions that can meet all typical interface print/apply requirements or be compatible with the functions of all major competitors' interfaces.

When combined with the GPIO Manager software, the GPIO module leverages Printronix's exclusive solutions enablers, such as the Online Data Validation (ODV™) system, resulting in the industry's most cost-effective solution, using the least amount of hardware.

The GPIO Manager software is a flexible, PC-based program offering simple and intuitive programming tools, including a graphical user interface (GUI) with pull-down menus, "self-documenting" with an audit trail for each line, and the ability to edit and upgrade quickly on a single screen. Additionally, the GPIO Manager software can program and map desired actions from the T5000^e series printer and GPIO accessory module to pre-defined events, such as "printer online", "label present", "label taken" and "printer error".

The GPIO accessory module contains eight optically isolated inputs, eight optically isolated outputs and four relay outputs, all available for GPIO Manager mapping. In addition, printer functions, such as keypad switches, serial communication ports and proprietary ODV analysis parameters, are available with the GPIO Manager, extending its mapping functions.

Integrators of the T5000^e series printers with label application systems can program a seamless system with simultaneous functions when a label is printed. For example, signals are sent to the label applicator, a light stack is illuminated for visual bar code or printer status indications, and printing

resumes after the label applicator has received the label. Additionally, printer keys can be programmed with special functions for custom user access, and specially formatted priority data transmissions can be sent to a host for specific conditions. The system allows virtually boundless creative and economic solutions for value-add opportunities.

PROFESSIONAL SERVICES

Your environment can be simple or complex. Your printers can be connected directly to your host, or scattered on networks around the globe. Your production line might have verifiers validating printed bar code labels as they move down a conveyor. No matter how your environment is set up, leveraging the full capability of your printers and verifiers may require consultation, development, customization, or training to ensure maximum productivity.

Printronix Professional Services can deliver what you need.

Many companies require their trading partners to adhere to strict guidelines and standards as integral components of the automated supply chain, or else face the prospects of returned goods or even fines. Printronix has the proven experience to handle retailer and vendor supply chain requirements, freeing your organization to focus on its core business.

It's not just about sending print jobs. It's about global printer management - error detection, notification and correction; remote troubleshooting; remote software updates; and more. It's not just about verifying bar codes on a label. It's about how your operation can run non-stop with full assurance that critical steps in your process, which depend on successful bar code scanning, will take place without fail.

Auto-ID/RFID Consulting

As a member of EPCglobal, Printronix Professional Services can provide a full range of expertise, from bar coding to RFID. Our specialists can provide on-site assessments, training and integration of bar code and RFID systems. We can guide you through a rapid implementation of smart label deployment, and assist in developing and implementing a successful migration strategy from bar code to RFID.

Closed-Loop Bar Code Verification

Printronix Professional Services can provide customers a full auditing of every label and bar code printed. Our specialists can provide integration to external corporate databases, including Oracle, DB2, SQL/Server, MySQL and others. We can develop custom management reports to address compliance, charge-backs and track quality control metrics, and create custom software to identify erroneous bar code content such as duplicate or out of sequence serial numbers.

Label Compliance and Certification

Using our expertise and close association with industry leaders, we can assess root cause for failed labels and bar codes against any automotive, retail or pharmaceutical specification. Printronix Professional Services is a GM and Sears approved certification provider. We can provide assessment and solutions for nearly all label related charge back fines. We can implement a complete auditing and quality control system to record label and bar code quality history, and can create compliance labels to meet any specification.

SAP and ERP Bar Code Integration and Implementation

Printronix Professional Services can facilitate the integration of SAP/ERP printing through custom device types and drivers. We offer ABAP/4 and SAP script support and are experts in all available middleware packages.

Legacy Application Migration and Implementation

We can migrate data streams to new printing platforms without modifying host applications. We offer services to redesign and rewrite applications for new environments. We can apply an array of software tools to bridge the translation, as well as provide Microsoft Back Office integration.

For the location of your nearest Printronix representative, call 800-826-3874.

Printronix, Inc.
14600 Myford Rd.
P.O. Box 19559
Irvine, CA 92623-9559

www.printronix.com

SINGAPORE
ASIA REGIONAL SALES OFFICE
Printronix Schweiz GmbH
No. 42, Changi South Street 1,
Changi South Industrial Estate
Singapore 486763
Tel: (65) 6542-0110 Fax: (65) 6546-1588

AUSTRALIA
AUSTRALIA/NEW ZEALAND OFFICE
Printronix Australia Pty. Ltd
Level 21, 201 Miller Street,
North Sydney NSW 2060
Tel: (61-2) 99592250   Fax: (61-2) 99592244

CHINA
SHENZHEN REGIONAL SALES OFFICE
Printronix Printer (Shenzhen) Co. Ltd
Unit F, 17/F Shenzhen International Trade Commercial Building
Nan Hu Road,
Luo Hu District, Shenzhen China 518014
Tel: (86-755) 25194027   Fax: (86-755) 25194019

BEIJING SALES OFFICE
Printronix Beijing
Room 1831, 18F, China Merchants Tower
No 118 Jian Guo Road
Chao Yang District
Beijing 100022 PR China
Tel: (86-10) 65662731   Fax: (86-10) 65662730

SHANGHAI OFFICE
Printronix Shanghai
31F, Jin Mao Tower,
88 Shi Ji Avenue
Pudong Shanghai
200120 PR China
Tel: (86-21) 2890 9719   Fax: (86-21) 2890 9219

INDIA
ASIA PTE REGIONAL SALES OFFICE
Printronix Asia Pte Ltd
B202 2nd Floor, Crystal Plaza,
Link Road, Andheri (W)
Mumbai 400053, India
Tel: (91-22) 26733423
Fax: (91-22) 26733422

KOREA

Printronix Asia Pte Ltd Korea Liaison Office
#3001, 30th Floor, Trade Tower,
159-1, Samseong-dong, Gangnam-gu
Seoul 135-729, Korea
Tel: (82-2) 60072066
Fax: (82-2) 60072723

EUROPE

Printronix Europe, Middle East and Africa Sales and Marketing
Headquarters:
PRINTRONIX FRANCE Sarl
8, rue Parmentier
F-92800 Puteaux
France
Tel: +33 (0) 1 46 25 19 00
Fax: +33 (0) 1 46 24 19 19
Email: EMEAsales@printronix.com

Index

About the Authors

Bob Kleist was a founder of Printronix Inc. in 1974 and currently serves as President, CEO and a director of the Irvine, California headquartered global company that designs, manufactures and markets a range of computer printers for business and industrial applications. Previous experience included a co-founder and executive of Pertec Computer Corporation and engineering and management assignments at Ampex, Link Aviation, and Magnavox. Bob holds 17 patents for peripheral and control systems and received a BSEE degree from Kansas University and an MSEE degree from Stanford University. He has served on the Seagate Technology board of directors, the Stanford Engineering Advisory Committee and actively supports engineering and computer science education at Stanford University and the University of California, Irvine.

Theodore A. Chapman is the senior vice president of engineering and product marketing and chief technology officer of Printronix. Chapman joined Printronix in November 1995 as vice president, product development. In April 1999, he was appointed as senior vice president, engineering and chief technical officer. From July 1970 to October 1995, Chapman held various engineering and senior management positions with IBM Corp.

David Sakai is the vice president of marketing at Printronix. Before joining Printronix, Sakai served as vice president of channel operations for Xerox Corp.'s Office Solutions Group. During his 24-year tenure at Xerox, Sakai held numerous senior management positions throughout the organization including general manager of California coastal operations; vice president of marketing, light lens business unit; and vice president of marketing, office systems group.

Brad Jarvis joined Printronix in April 2003, as director of product marketing. Jarvis has lectured extensively and has been quoted as a recognized expert on the subject of RFID in publications around the world. Prior to joining Printronix, he was Director of Products and Markets for Lantronix, a hardware and software solutions provider. Jarvis also spent 14 years with Xerox in a range of senior management positions in sales and marketing.